D1488343

PHOTOELASTIC COATINGS

SOCIETY FOR EXPERIMENTAL STRESS ANALYSIS MONOGRAPH SERIES

PHOTOELASTIC COATINGS

Felix Zandman
Vishay Intertechnology, Inc. 644-1300
Malvern, Pennsylvania

Salomon Redner
Photolastic, Inc.
Malvern, Pennsylvania

James W. Dally
University of Maryland
College Park, Maryland

─────────────Published Jointly by─────────────
THE IOWA STATE UNIVERSITY PRESS
Ames, Iowa
SOCIETY FOR EXPERIMENTAL STRESS ANALYSIS
Westport, Connecticut

SESA Monograph No. 3
This monograph is published in furtherance of SESA objectives in the field of experimental mechanics. The Society is not responsible for any statements made or opinions expressed in its publications.

Library of Congress Cataloging in Publication Data

Zandman, Félix.
    Photoelastic coatings.

    (SESA monograph; no. 3)
    Bibliography: p.
    Includes indexes.
    1. Photoelasticity. 2. Plastic coating.
I. Redner, Salomon, 1929– joint author.
II. Dally, James W., joint author. III. Title.
IV. Series: Society for Experimental Stress
Analysis. SESA monograph; no. 3.
TA418.12.Z36   620.1'123   76-46984
ISBN 0-8138-0035-8

First edition, 1977

CONTENTS

PREFACE

This book is the third in a series of volumes published by the Monograph Committee of the Society for Experimental Stress Analysis. The Society feels that, in addition to its primary dissemination of information through publication of the journal *Experimental Mechanics*, this series of monographs is a vital means of providing the engineer with a handy and in-depth source in a specific field.

Photoelastic Coatings is written by three highly qualified authorities in the field, Felix Zandman, Salomon Redner, and James W. Dally. The material contained in this volume represents not only more than twenty years experience for each author but also research of the numerous other workers that are cited. Dr. Zandman, in particular, has extended considerable effort and devotion to ensure acceptance of and improvements in techniques involving photoelastic coatings. This commitment has resulted in lectures that have literally spanned the globe. In 1970 the SESA awarded Dr. Zandman the Distinguished Contribution Award with the citation: "For outstanding original contributions to the technology of experimental mechanics through the development of birefringent coatings." Mr. Redner's contributions range from instrument and coating developments to improvement of methods. Professor Dally's leadership in the application of the concept of photoelasticity to a wide variety of problems in experimental mechanics is well documented. The Society is fortunate to have the services of these specialists to write on this important field.

The first two chapters of this monograph are devoted to background material. Chapter 1 deals with the general theory of photoelasticity; the specific application of this theory to photoelastic coatings is covered in Chapter 2. The next two chapters discuss the materials used for coatings and the equipment employed to analyze the photoelastic patterns. In the chapters that follow, the parameters that must be considered in various applications and the specific applications themselves are discussed. The aim of this monograph is to serve as a quick reference for those already in the field; and, with the background material presented, to aid the practicing engineer in be-coming proficient in this area with a minimum effort.

The Monograph Committee wishes to thank the authors who voluntarily gave their time to produce this volume. Their coopera-tion and enthusiasm is greatly appreciated. As usual, B. E. Rossi, Managing Director of the Society for Experimental Stress Analysis, displayed his untiring effort in assuming responsibility for editorial styling and working with the publishers.

The SESA Committee on Monographs

M. E. Fourney, Chairman
P. H. Adams
J. W. Dally
A. S. Kobayashi
C. E. Taylor

INTRODUCTION

Many experimental methods can be used to determine the distribution of stresses and strains in machine components and structures. Probably the most common method involves the use of resistance strain gages that are easy to install and provide precise results under a wide variety of operating conditions. Unfortunately, strain gages provide data only at the points where they have been mounted, and little is learned about other regions on the surface of the component. As a consequence, there has been a continuing interest in "full-field" experimental methods that provide data on stresses and strains over relatively large areas of the component surface.

Brittle coatings or lacquers represent one of the full-field experimental methods. These coatings are similar to a varnish except that they crack when subjected to a critical strain. The cracking of the coating is useful in locating the position and direction of the maximum stress. However, since the resin-based brittle coatings are extremely sensitive to a number of parameters such as humidity, temperature, loading time, and coating thickness, precise determination of the stress magnitude is not always possible.

Two- and three-dimensional photoelasticity also provides full-field experimental methods to determine stresses at points on the surface or in the interior of the specimen. Precise results can be obtained, although the three-dimensional method is more complex than the strain-gage and brittle-coating methods. Classical photoelasticity requires the use of models fabricated from transparent birefringent polymers and this also limits the applicability of the method.

The method of birefringent coatings, also called the method of photoelastic coatings, extends the classical procedures of model photoelasticity to the measurement of surface strains in opaque two- and three-dimensional models made of any structural material. The coating is a thin layer of birefringent material—usually a polymer—that is bonded integrally to the flat or curved surfaces of the prototype being analyzed for stress. When the prototype is loaded, the surface strains are transmitted to the coating, reproducing the prototype strain field in the coating. (For certain cases in which the prototype strain field is affected by the presence of the coating, coating-thickness correction factors are introduced to account for this effect.) To provide light reflection at the interface, the coating is bonded in place with a reflective cement. When viewed through a white-light reflection polariscope, the strained coating exhibits black isoclinic and colored isochromatic fringes. Isoclinic fringes provide directions of principal strains. Isochromatic fringes, when viewed in normal-incidence light, provide the difference of principal strains (maximum shear strains); when viewed in oblique-incidence light, they provide additional data that permit the determination of the magnitude and sign of individual principal strains. Full-field isochromatic and iso-clinic patterns are directly visible and can be photographed for subsequent analysis by using simple reflection polariscopes. With more advanced equipment, the fringe patterns can be analyzed on a point-by-point basis by using a compensator to determine the fringe order, with the magnitude of the strain being displayed on a digital readout device.

The photoelastic-coating method has many advantages compared to other methods of experimental stress analysis. It provides point-by-point or full-field quantitative data, enabling the investigator to determine the complete distribution of surface strains and directly highlighting severely strained areas. The method is nondestructive and, since the coatings can be applied directly to prototype parts and structures, eliminates the need for elaborate and costly models. Both static and dynamic strains can be measured. With appropriately selected coating materials, the method is applicable over a wide range of elastic and plastic strains. The photoelastic-coating method is also very useful in converting analysis of complex nonlinear stress situations in the prototype to analysis of relatively simple linear-elastic problems in the coating; i.e., plastic and viscoelastic behavior in a prototype can be measured in terms of the elastic response of the coating. Similarly, the anisotropic characteristics of composite materials can be examined in terms of an isotropic response in the coating.

The concept of photoelastic coatings was first proposed by M. Mesnager[1] in France in 1930, and the method was reexamined by Oppel[2] in Germany in 1937. Mesnager tried to bond segments of glass to structures. The difficulties in machining glass; its high modulus, which caused significant reinforcement; and the lack of a proper adhesive for bonding the structure prevented this approach from being developed. Oppel used flat sheets of Bakelite. Here, the development of a severe "time-edge effect," lack of strong adhesives, and applicability only to flat surfaces prevented this method from being used industrially.

With the availability of epoxy resins in the 1950s, the required high-strength adhesive and a photoelastic sheet relatively free of time-edge effect could be produced. The development of the photoelastic-coating method proceeded rapidly with contributions in materials, techniques, and instruments from Fleury and Zandman[3] in France; D'Agostino, Drucker, and Liu[4,5] in the United States; and Kawata[6] in Japan.

At this time photoelastic coatings were still considered an academic curiosity by those in industry because a technique to apply coating surfaces with compound curvature had not been developed. Finally, Zandman[7] developed the contour sheet procedure for applying a constant-thickness coating to curved surfaces without introducing residual birefringence. At the same time he developed a portable reflection polariscope,[8] making the method practicable for quantitative measurements under industrial conditions. In the late 1950s Zandman, Redner, and Riegner[9] treated the problem of reinforcement by the coating and developed techniques to account for reinforcing effects, thus removing an obstacle to obtaining quantitative data when coating thickness is not negligible with respect to the thickness of the part.

Following these developments of epoxy coating materials and adhesives, sheet-contouring techniques, and analysis procedures, the method was generally accepted and considerable research followed. Investigators developed more sensitive and higher elongation coatings. Several instruments for laboratory and field use, particularly designed for coatings, were introduced. Methods were advanced for using the coatings in analysis of plasticity, thermoelasticity, vibrations, and wave and crack propagation. Also, many engineers in industry began to use the method in the solution of a wide variety of industrial design problems.

No attempt is made in this monograph to recite the individual contributions of the many investigators who improved the method. Extensive literature on this subject has appeared in the United States, France, United Kingdom, U.S.S.R., and other countries (see References). Instead, this volume is arranged to guide the experimentalist in the use of birefringent coatings toward the solution of practical stress-analysis problems. The background necessary to thoroughly understand the theory of the photoelastic-coating method is presented in Chapters 1 and 2. Materials commonly used for coatings and adhesives, along with the techniques and procedures for bonding the coating to the prototype, are described in Chapter 3. Reflection polariscopes, compensators, and calibration devices adapted for use with birefringent coatings are described in Chapter 4. Parameters affecting the behavior of the coating and correction factors for coating thickness and other effects are treated in Chapter 5. A series of case histories is presented in Chapter 6 to illustrate the application of the coating method to components fabricated from different materials and to different types of problems that occur in a wide range of industries. Finally, Chapter 7 discusses likely future developments.

DEFINITION OF SYMBOLS

β	= angular position of quarter-wave plates
δ	= linear retardation
Δ	= angular retardation
$\epsilon_1, \epsilon_2, \epsilon_3$	= principal strains
$\epsilon_{xx}, \epsilon_{yy}, \epsilon_{zz}$	= normal strains
γ	= rotation angle of the analyzer
$\gamma_{xy}, \gamma_{xz}, \gamma_{yz}$	= shear strain
ϕ, φ	= isoclinic parameter, direction of principal strains
ν	= Poisson's ratio
θ	= angle of incidence of polarized light
λ	= wavelength of light
ρ	= radius of curvature
Γ	= constants of integration
σ_1, σ_2	= principal stresses
$\sigma_{xx}, \sigma_{yy}, \sigma_{zz}$	= normal stresses
$\tau_{xy}, \tau_{xz}, \tau_{yz}$	= shear stress
ω	= circular frequency of light
a	= maximum amplitude
c	= speed of light
f	= frequency
f_ϵ, f_σ	= material-fringe value (strain and stress)
g	= thickness ratio of plastic coating to metal, h_c/h_s
h	= thickness
n, n_0, n_1, n_2	= index of refraction
r	= fractional fringe order
t	= time
u	= displacement
z	= position along the axis of propagation
A	= amplitude of light vector components
A_1, A_2, A_3	= light vector components
C	= stress-optic coefficient
$C_{1,2,3,4,5}$	= correction factors
E	= modulus of elasticity
F_ϵ, F_σ	= coating fringe value
I	= intensity of light
K	= strain-optic sensitivity of the coating
N, N_0, N_θ	= fringe order (relative retardation)
S_σ	= stress-sensitivity index
T	= period

SUMMARY OF BASIC EQUATIONS

1. UNIAXIAL-STRESS MEASUREMENTS ($\sigma_2 = 0$)

Measured: Fringe order in normal incidence, N
  Isoclinic angle, ϕ

Coating: Thickness h, sensitivity K, fringe constant $F_\epsilon = \lambda/2hK$

Difference of principal strains: $\epsilon_1 - \epsilon_2 = NF_\epsilon$

Strains: $\epsilon_1 = N[F_\epsilon/(1+\nu)]$; $\epsilon_2 = -\nu\epsilon_1 = -N[F_\epsilon\nu/(1+\nu)]$

Stress: $\sigma_1 = E\epsilon_1 = N[EF_\epsilon/(1+\nu)]$

2. BIAXIAL-STRESS MEASUREMENTS

Measured: Fringe order in normal incidence, N_0
  Fringe order in oblique incidence, N_θ
  Isoclinic angle, ϕ

Coating: Thickness h, sensitivity K, fringe constant $F_\epsilon = \lambda/2hK$

Difference of principal strains: $\epsilon_1 - \epsilon_2 = N_0 F_\epsilon$

Shear strain: $\gamma_{xy} = N_0 F_\epsilon \sin 2\phi$

Separated values of strain:

$$\epsilon_x = [F_\epsilon/(1+\nu_c)\sin^2\theta_x][N_\theta(1-\nu_c)\cos\theta_x - N_0(\cos^2\theta_x - \nu_c)]$$

$$\epsilon_y = [F_\epsilon/(1+\nu_c)\sin^2\theta_x]\{N_\theta(1+\nu_c)\cos\theta_x - N_0[(1-\nu_c)\cos^2\theta_x]\}$$

$$N_\theta = [1/F_\epsilon(1-\nu_c)\cos\theta]\{\epsilon_x(1-\nu_c\cos^2\theta) - \epsilon_y(\cos^2\theta - \nu_c)\}$$

Difference of principal stresses: $\sigma_1 - \sigma_2 = N_0[EF_\epsilon/(1+\nu_c)]$

Separated values of stresses: $\sigma_{1,2} = [E/(1-\nu^2)](\epsilon_{1,2} + \nu\epsilon_{2,1})$

Shear stress: $\tau_{xy} = [NF_\epsilon E/2(1+\nu)]\sin 2\phi$

3. CORRECTION FACTORS

To obtain surface strains on uncoated structure ϵ^u from the measured strains in the coating ϵ^c:

$$\epsilon_1^u - \epsilon_2^u = (1/C_n)(\epsilon_1^c - \epsilon_2^c)$$

C_n: correction factors shown on Figs. 5.2, 5.4, 5.5, and 5.6.

PHOTOELASTIC COATINGS

SOCIETY FOR EXPERIMENTAL STRESS ANALYSIS MONOGRAPH SERIES

1
PROPERTIES OF LIGHT AND ELEMENTARY THEORY OF PHOTOELASTICITY

1.1 BEHAVIOR OF LIGHT

The photoelastic effect can be adequately described by the electromagnetic theory of light propagation. This theory states that light is an electromagnetic disturbance, which may be represented by a light vector that is normal to the direction of propagation. In ordinary light as produced by a tungsten-filament bulb, the light vector is randomly oriented and may be described by a number of arbitrary vibrations, as illustrated in Fig. 1.1.

The disturbance-producing light may be described as a wave motion where the amplitude A of the light vector is given by

$$A = a \sin (2\pi/\lambda) (z - ct) \tag{1.1}$$

where z = position along the axis of propagation
t = time
c = velocity of propagation (3×10^8 m/s in vacuum)
λ = wavelength
a = maximum amplitude

A graphical representation of the amplitude of the light vector as it propagates along the z axis is given in Fig. 1.2. The wavelength λ is the distance between peaks and is related to the period or time required for passage of two successive peaks at some fixed station z. The period T is given by

$$T = \lambda/c \tag{1.2}$$

The frequency of the light vector is the number of oscillations per

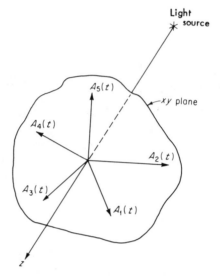

Fig. 1.1—Components of the transverse vibrations that produce the random-light vector associated with ordinary light.

second; thus the frequency f is given by

$$f = 1/T \tag{1.3}$$

The color of light observed by the eye is determined by the frequency f. The colors in the visible spectrum range from deep red with a frequency of 390×10^{12} Hz to a deep violet with a frequency of 770×10^{12} Hz. Although most photoelastic investigations are conducted using visible light, the principles of photoelasticity are valid into the infrared and the ultraviolet regions of radiant energy.

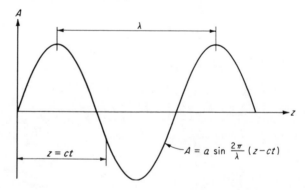

Fig. 1.2—Amplitude of the light vector as a function of position along the axis of propagation.

When the light vector is composed of several components (A_1, A_2, A_3, etc.) all having the same frequency, the light is classified as monochromatic, its color depending upon the frequency. However, the components A_1, A_2, A_3 of the light vector may exhibit different frequencies. In this instance the colors of the components are mixed, thus producing white light.

1.2 POLARIZED LIGHT

Ordinary light consists of random vector components that are transverse to the direction of propagation. When these components exhibit a preferential direction, the light is considered to be polarized. Three forms of polarization are encountered in photoelasticity. They are referred to as (1) linearly or plane polarized, (2) circularly polarized, and (3) elliptically polarized.

Plane-polarized light occurs when all components of the light vector lie in a single plane known as the plane of polarization. Circularly polarized light is obtained when the tip of the light vector describes a circular helix as the light propagates along the z axis. Elliptically polarized light is obtained when the tip of the light vector describes an elliptical helix as the light propagates along the z axis. The light vector producing plane, circularly, and elliptically polarized light is graphically represented in Fig. 1.3. In practice, plane-polarized light is produced with an optical filter known as a linear polarizer. Production of circularly or elliptically polarized light requires the use of a series combination of a linear polarizer and a retardation plate.

1.3 LINEAR POLARIZERS

Linear polarizers are optical elements that block the components of the light vector vibrating in the direction transverse to the axis of the polarizer. When a light vector passes through a linear polarizer (also called a plane polarizer), the component of the light vector that is perpendicular to the axis of polarization is blocked and the component parallel to the axis is transmitted as illustrated in Fig. 1.4.

Since the plane polarizer is at a fixed position along the z axis, the equation for the amplitude A given in eq. (1.1) may be expressed as

$$A = a \sin (2\pi/\lambda) \, ct = a \sin 2\pi ft = a \sin \omega t \qquad (1.4)$$

where $\omega = 2\pi f$ is the circular frequency of the light. The blocked and transmitted components of the light vector A_b and A_t respec-

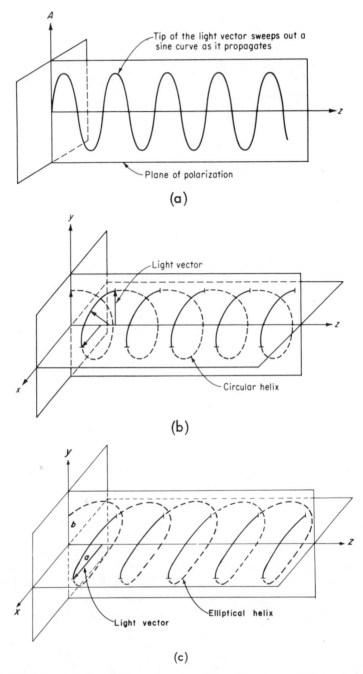

Fig. 1.3—Motion of the light vector describing (a) plane, (b) circularly, and (c) elliptically polarized light.

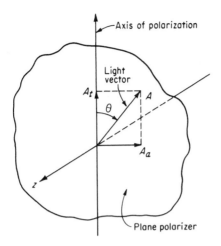

Fig. 1.4—Absorbing and transmitting characteristics of a linear polarizer.

tively are given by

$$A_b = a \sin \omega t \sin \theta \qquad A_t = a \sin \omega t \cos \theta \qquad (1.5)$$

where θ is the angle between the light vector A and the axis of polarization shown in Fig. 1.4. Plane-polarized light is usually produced with polarizing filters that provide a large field of very well polarized light at a relatively low cost.

1.4 RETARDATION PLATES

Certain transparent materials (e.g., crystals and stressed plastics) have the ability to resolve an impinging light vector into two orthogonal components and to transmit each with a different velocity. Transparent materials exhibiting this property are known as doubly refracting. A doubly refracting plate has two principal axes, noted as 1 and 2 in Fig. 1.5. The propagation of light along axis 1 proceeds at a velocity c_1, and along axis 2 at a velocity c_2. Since $c_1 > c_2$, axis 1 is often called the fast axis and axis 2, the slow axis.

If this doubly refracting plate is placed in a field of linear polarized light such that the light vector A_t makes an angle β with axis 1, the light vector entering the plate is resolved into two components A_{t1} and A_{t2} along axes 1 and 2 respectively, with

$$A_{t1} = A_t \cos \beta = a \cos \theta \sin \omega t \cos \beta = k \sin \omega t \cos \beta$$

$$A_{t2} = A_t \sin \beta = a \cos \theta \sin \omega t \sin \beta = k \sin \omega t \sin \beta \qquad (1.6)$$

where $k = a \cos \theta$.

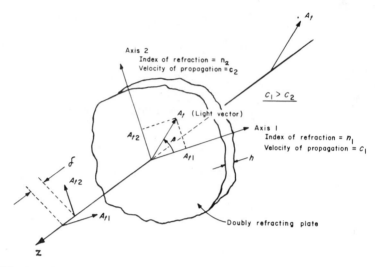

Fig. 1.5—Plane-polarized-light vector entering a doubly refracting plate.

The light components A_{t1} and A_{t2} propagate through the plates with velocities c_1 and c_2; and since $c_1 > c_2$, the A_{t2} component is retarded relative to the A_{t1} component. This retardation can be visualized by considering the relative phase shift between the two components as shown in Fig. 1.6. The absolute retardation δ of each component is

$$\delta_1 = h(n_1 - n) \qquad \delta_2 = h(n_2 - n) \qquad (1.7)$$

where h = thickness of the plate

n = index of refraction of air

n_1, n_2 = indices of refraction associated with axes 1 and 2

If c is the velocity in air, $n_{1,2} = c_{1,2}/c$. The difference $\delta_1 - \delta_2$ represents the relative retardation

$$\delta = \delta_1 - \delta_2 = h(n_1 - n_2) \qquad (1.8)$$

The angular phase shift between the two components of light, as shown in Fig. 1.5, is given by

$$\Delta = (2\pi/\lambda)\delta = (2\pi h/\lambda)(n_1 - n_2) \qquad (1.9)$$

Thus the angular retardation Δ produced by a retardation plate is dependent on its thickness h, the wavelength of the light λ, and the properties of the plate described by $n_1 - n_2$. When a doubly refracting plate is designed with $\delta = \lambda/4$ ($\Delta = \pi/2$), it is called a quarter-wave plate. Similarly, doubly refracting plates with $\delta = \lambda/2$ and $\delta = \lambda$ are known as half-wave and full-wave plates respectively.

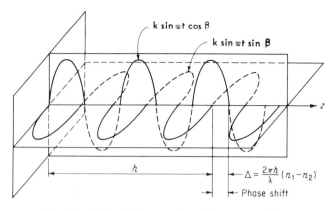

Fig. 1.6—Retardation produced by a wave plate.

Upon passage of the light through a doubly refracting plate with an angular retardation of Δ, the equations describing the amplitudes of the two emerging components are

$$A'_{t1} = k \cos \beta \sin (\omega t + \Delta) \qquad A'_{t2} = k \sin \beta \sin \omega t \qquad (1.10)$$

The amplitude of the light vector produced by the resultant of the two mutually perpendicular components is

$$A'_t = [A'^2_{t1} + A'^2_{t2}]^{1/2}$$

$$= k [\sin^2 (\omega t + \Delta) \cos^2 \beta + \sin^2 \omega t \sin^2 \beta]^{1/2} \qquad (1.11)$$

and the angle γ that the emerging light vector makes with axis 1 is given by

$$\tan \gamma = A'_{t2}/A'_{t1} = [\sin \omega t/\sin (\omega t + \Delta)] \tan \beta \qquad (1.12)$$

These results show that both the amplitude and the angle of rotation of the emerging light vector can be controlled by the retardation plate. The characteristics of the emerging light vector depend upon the angular retardation of the wave plate Δ and the orientation angle β. Quarter-wave plates are usually fabricated from thin polymeric sheets that have been permanently stretched to provide the molecular orientation, thus making them doubly refracting.

1.5 CONDITIONING LIGHT WITH A LINEAR POLARIZER AND A RETARDATION PLATE

The equations for the amplitude and direction of the light vector emerging from a series combination of a linear polarizer and retardation plate are given in eqs. (1.11) and (1.12). The light emerging from this combination is always polarized; however, the type of polarization may be plane, circular, or elliptical. The selection of the angular retardation Δ of the retardation plate and the orientation angle β determines the type of polarized light that emerges. Three cases are of fundamental importance.

Case 1—Plane or Linearly Polarized Light

If the angle $\beta = 0$ and Δ is any value, the amplitude and direction of the emerging light vector are determined from eqs. (1.11) and (1.12) as

$$A_t' = k \sin(\omega t + \Delta) \qquad \gamma = 0 \qquad (1.13)$$

Since $\gamma = 0$, the light vector is not rotated as it passes through the retardation plate and the emerging light is still plane polarized. The retardation plate in this case does not influence the light except to retard it by an angle Δ. The same results are also obtained with $\beta = \pi/2$ where the axis of the linear polarizer is aligned with the slow axis of the quarter-wave plate.

Case 2—Circularly Polarized Light

If a retardation plate is selected with $\Delta = \pi/2$ (a quarter-wave plate) and $\beta = \pi/4$, the emerging light vector is given by

$$A_t' = (\sqrt{2}/2)k (\sin^2 \omega t + \cos^2 \omega t)^{1/2} = (\sqrt{2}/2)k$$

$$\gamma = \omega t \qquad (1.14)$$

These results indicate that the emerging light vector exhibits a constant amplitude and that the angle of emergence increases as a linear function of time t. It is clear then that the tip of the light vector sweeps out a circle; and as the light propagates down the z axis, the circle is expanded into a circular helix.

Case 3—Elliptically Polarized Light

If a quarter-wave plate with $\Delta = \pi/2$ is selected and β is any angle other than 0, $\pi/4$, or even multiples thereof, from eqs. (1.11) and (1.12) the emerging light vector is described by

$$A_t' = k [\cos^2 \omega t \cos^2 \beta + \sin^2 \omega t \sin^2 \beta]^{1/2}$$

$$\tan \gamma = \tan \omega t \tan \beta \qquad (1.15)$$

By letting $A_{t1}' = x$, $A_{t2}' = y$, $k^2 \cos^2 \beta = a^2$, and $k^2 \sin^2 \beta = b^2$, it is possible to show that the tip of the light vector sweeps out an ellipse in the x-y plane and that eq. (1.15) can be expressed as $(x/a)^2 +$ $(y/b)^2 = 1$. As the light propagates along the z axis, the ellipse is expanded into an elliptical helix.

1.6 ARRANGEMENT OF THE OPTICAL ELEMENTS IN A POLARISCOPE

Two different optical arrangements are commonly employed in photoelasticity to provide the optical systems known as the plane polariscope and the circular polariscope. The plane-transmission polariscope is the simplest system since it consists of only two linear polarizers and a light source arranged as shown in Fig. 1.7(a). The linear polarizer adjacent to the light source is called the polarizer while the linear polarizer next to the observer is known as the analyzer. In the plane polariscope the two axes of polarization are always crossed; thus no light is transmitted through the analyzer, and this optical system produces a dark field. When using the plane polariscope, the stressed photoelastic material (model or coating) is inserted between the two crossed elements and viewed through the analyzer. Two superimposed fringe patterns are observed—the isoclinic and the isochromatic; however, a discussion of these patterns will be deferred until Section 1.8.

When using birefringent coatings, the light must be reflected from the interface between the coating and the prototype as shown in Fig. 1.7(b). The use of reflected light is accomplished by folding the optical axis with a partial mirror. Further details on plane polariscopes for birefringent coatings are given in Chapter 4.

The circular polariscope, as the name implies, employs circularly polarized light so that the optical system incorporates series combinations of linear polarizers and quarter-wave plates as illustrated in Fig. 1.8(a) for the transmission polariscope. The first polarizer and the quarter-wave plate are arranged with $\beta = \pi/4$ to produce circularly polarized light. The second combination is arranged in reverse order and is usually set crossed to the first elements. This arrangement of the four elements with the polarizer and analyzer crossed and the quarter-wave plates crossed produces a dark field since no light is transmitted through the second polarizer. Actually, four arrangements of the optical elements in the circular polariscope are possible depending upon whether the polarizers and quarter-wave plates are crossed or parallel. These four optical arrangements are described in Table 1.1. A more complete description of the circular polariscope is given in Ref. 8.

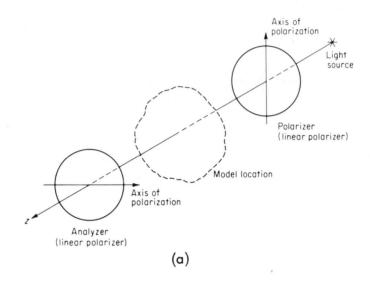

(a)

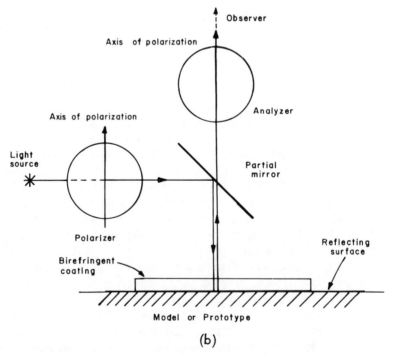

(b)

Fig. 1.7—(a) Plane transmission polariscope, and (b) plane reflection polariscope.

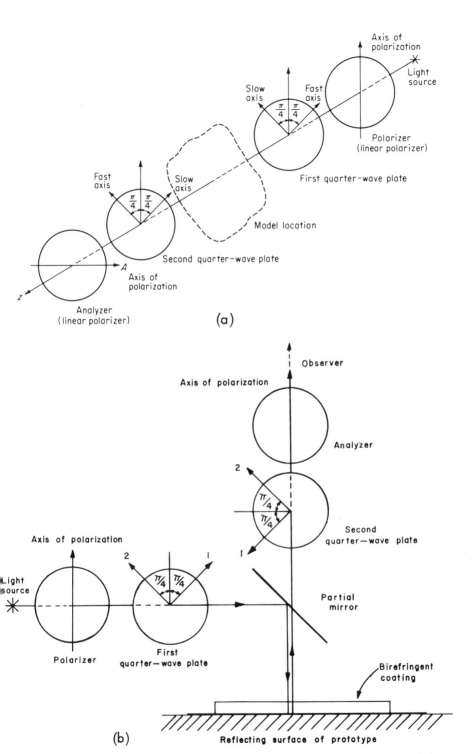

Fig. 1.8—(a) Circular transmission polariscope, and (b) circular reflection polariscope.

Table 1.1—Arrangements of the Optical Elements in a Circular Polariscope

Arrangement	Quarter-wave Plates	Polarizer and Analyzer	Field
A	Crossed	Crossed	Dark
B	Crossed	Parallel	Light
C	Parallel	Crossed	Dark
D	Parallel	Parallel	Light

1.7 THE STRESS-OPTIC LAW

When a photoelastic model is placed in a polariscope and subjected to either a state of stress or strain, isoclinic and/or isochromatic fringe patterns are observed. To interpret these fringe patterns, it is necessary to determine the relation between the applied stress or strain and the optical effects observed in the polariscope.

Consider a photoelastic model machined from a sheet of a suitable polymer as an optical element. Initially, the model is stress-free and exhibits a constant index of refraction n_0 at all points. When a system of loads is placed on the model, a plane state of stress is developed that changes the optical properties of the model material. The model becomes optically anisotropic and acts as a doubly refracting plate, as discussed in Section 1.4. The principal axes of stress coincide with the principal optical axes and the indices of refraction depend upon the magnitude of the stresses at each point in the plate. It is this property of the model to act as a "temporary" retardation plate when subjected to stress that serves as the basis for photoelasticity. The polariscope is simply the optical instrument used to measure the changes that occur in the index of refraction.

The relation between the stresses and the index of refraction was formulated by Maxwell in 1890 as

$$n_1 - n_0 = C_1 \sigma_1 + C_2 \sigma_2 \qquad n_2 - n_0 = C_1 \sigma_2 + C_2 \sigma_1$$

where n_0 = the index of refraction of the model or coating in the unstressed state

n_1, n_2 = the indices of refraction along the two principal axes associated with σ_1 and σ_2 respectively

C_1, C_2 = the stress-optic coefficients

Subtracting to eliminate n_0 gives

$$n_1 - n_2 = (C_1 - C_2)(\sigma_1 - \sigma_2) = C(\sigma_1 - \sigma_2) \qquad (1.16)$$

where $C = C_1 - C_2$ is the relative stress-optic coefficient.

Recall next that the model is acting as a temporary retardation plate and eqs. (1.8) and (1.9) indicate that the relative and angular retardation is related to the $n_1 - n_2$ as

$$\delta/h = n_1 - n_2 \qquad \Delta\lambda/2\pi \, h = n_1 - n_2 \qquad (1.17)$$

Substituting eqs. (1.17) in eq. (1.16) gives

$$\delta = h \, C \, (\sigma_1 - \sigma_2)$$

$$\Delta = (2\pi/\lambda) \, h \, C \, (\sigma_1 - \sigma_2) \qquad (1.18)$$

Equations (1.18) are the classical description of the stress-optic law. The relative angular retardation Δ is linearly proportional to the difference in the principal stresses $\sigma_1 - \sigma_2$. The relative stress-optic coefficient C is a property of the polymeric material used to fabricate the model and is usually treated as a constant. The relative retardation is linearly related to the thickness of the plate and inversely related to the wavelength of light employed with the polariscope.

The stress-optic law is more commonly rewritten today as

$$\sigma_1 - \sigma_2 = N f_\sigma/h \qquad (1.19)$$

where $N = \Delta/2\pi = \delta/\lambda$ = the relative retardation in terms of a complete cycle, also called "fringe order"
$f_\sigma = \lambda/C$ = the material fringe value

It is apparent from eq. (1.19) that the principal-stress difference can be determined if the material fringe value of the model material can be established by a calibration procedure and if N can be measured at each point by observation in a polariscope. With birefringent coatings, the light traverses the model twice, so the form of the stress-optic law is modified as

$$\sigma_1^c - \sigma_2^c = N f_\sigma/2h_c = N F_\sigma \qquad (1.20)$$

where $F_\sigma = f_\sigma/2h_c$ is the coating fringe value, and c is used as a superscript and a subscript to refer to application of the stress-optic law to photoelastic coatings.

Photoelastic models can also be used to determine the difference in the principal strains. From the two-dimensional form of Hooke's Law

$$\epsilon_1 - \epsilon_2 = [(1 + \nu)/E] \, (\sigma_1 - \sigma_2)$$

where ν = Poisson's ratio
E = the modulus of elasticity

Substituting this expression in eq. (1.20) gives

$$\epsilon_1^c - \epsilon_2^c = [(1 + \nu)/E] \, (f_\sigma/2h_c)N = (f_\epsilon/2h)N = F_\epsilon N \qquad (1.21)$$

where $f_\epsilon = [(1 + \nu)/E]\,f_\sigma$, the material fringe value in terms of strain
$F_\epsilon = f_\epsilon/2h = \lambda/2hK$, the coating fringe value in terms of strain
K = the optical sensitivity constant of the coating (dimensionless)

Since birefringent coatings are usually used to determine surface strains on test components, eq. (1.21) is commonly employed. Reference should be made to Chapter 2 for further interpretation of the strain-optic law related to birefringent coatings.

1.8 EFFECTS OF A STRESSED PLATE IN A PLANE POLARISCOPE

Equations have been developed relating the principal-stress or principal-strain difference to the relative retardation N. Also, it has been stated that the optical axes of the model or coating coincide with the principal axes of stress. These two facts can be utilized in a stress analysis if a polariscope can be used for measuring N and the directions of the optical axes.

Consider first a plane-transmission polariscope where the model is positioned so the principal-stress direction σ_3 coincides with the z axis of the polariscope and that associated with σ_1 makes an angle α with the axis of the polarizer as shown in Fig. 1.9. Note that the light emerges from the polarizer and propagates to the model with an amplitude that may be expressed as

$$A = k \sin \omega t \tag{1.22}$$

The plane-polarized light enters the model or coating as shown in Fig. 1.10. Since the stressed model acts as a retardation plate, the

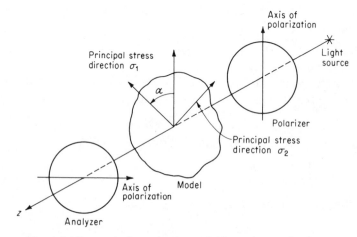

Fig. 1.9—Stressed photoelastic model in a plane polariscope.

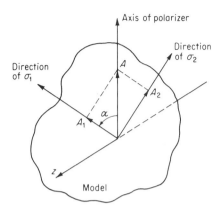

Fig. 1.10—Resolution of the light vector entering a stressed model in a plane polariscope.

impinging light vector is resolved into components A_1 and A_2 described by

$$A_1 = k \sin \omega t \cos \alpha \qquad A_2 = k \sin \omega t \sin \alpha \qquad (1.23)$$

These two components propagate through the stressed model at different velocities and are out of phase by an angular retardation Δ, which by eq. (1.19) is

$$\Delta = 2\pi N = 2\pi \ (h/f_\sigma) \ (\sigma_1 - \sigma_2) \qquad (1.24)$$

The amplitude of the light-vector components after emergence are

$$A_1' = k \cos \alpha \sin (\omega t + \Delta) \qquad A_2' = k \sin \alpha \sin (\omega t) \qquad (1.25)$$

Components A_1' and A_2' propagate out of phase by an angle Δ and impinge on the analyzer as shown in Fig. 1.11. The light components A_1' and A_2' are resolved into components A_1'' and A_2'' along the axis of the analyzer. Since the vertical components are perpendicular to the axis of the analyzer, they are blocked and only the horizontal components are transmitted. The light emerging from the analyzer is plane polarized with an amplitude given by

$$A = A_2'' - A_1'' = A_2' \cos \alpha - A_1' \sin \alpha \qquad (1.26)$$

Substituting eq. (1.25) into (1.26) and reducing with suitable triometric identities gives $A = -k \sin 2\alpha \cos \omega t \sin (\Delta/2)$.

The intensity of light as recorded by the eye or a sheet of film is proportional to the square of the amplitude of the light vector emerging from the analyzer. The intensity of light I corresponding to this

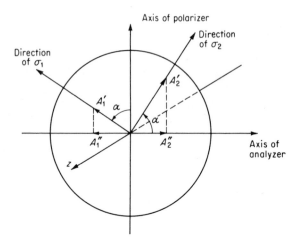

Fig. 1.11—Components of the light vector transmitted through the analyzer of a plane polariscope.

point on the model is

$$I = K_1 \sin^2 2\alpha \sin^2 (\Delta/2) \cos^2 \omega t \qquad (1.27)$$

where K_1 is a constant of proportionality. From the form of eq. (1.27) it can be seen that the intensity $I = 0$ and extinction can occur due to any one of three different conditions.

Case 1—Frequency of Light Effects

When $\omega t = (2n + 1)(\pi/2)$ with $n = 0$, 1, 2, etc., the term $\cos^2 \omega t = 0$; the intensity I goes to zero, producing a condition of extinction. However, the frequency is so high that extinction occurs on the order of 10^{13} times each second, so that the eye or any type of available high-speed photographic equipment does not record this periodic variation in the light intensity. The effect of this variation is averaged, and eq. (1.27) is rewritten as

$$I = K \sin^2 2\alpha \sin^2 (\Delta/2) \qquad (1.28)$$

where $K = (K_1/t^*) \int_0^{t^*} \cos^2 \omega t \, dt$, and t^* is the exposure time.

This frequency effect determines the frequency response for dynamic-stress applications of photoelasticity. Actually, the frequency response of the polariscope is quite high—on the order of 10^{12} Hz—because the frequency of the light can be taken as the carrier frequency in an analogy with an *a-c* amplifier.

Case 2—Effect of Principal-stress Directions

Referring again to eq. (1.27), the intensity I goes to zero and produces extinction when $2\alpha = n\pi$ and $\sin^2 2\alpha = 0$. Thus extinction occurs when $\alpha = 0$, $\pi/2$, and any integer multiple of $\pi/2$. Recalling the definition of α, this fact implies that the directions of the principal stresses coincide with the axes of the plane polariscope when an extinction occurs. Extending this analysis to all points in the model, extinction will be observed along lines formed by a locus of points where $\alpha = 0$ or $\pi/2$. These lines are the isoclinic fringes observed in a plane polariscope. In practice, several families of isoclinic fringes are determined by rotating the crossed polarizers P and A together relative to the fixed model and observing the isoclinic fringes after each increment of rotation.

Case 3—Effect of the Principal-stress Difference

When $\Delta/2 = n\pi$ with $n = 0$, 1, 2, etc., $\sin^2 (\Delta/2) = 0$ and the intensity I goes to zero, producing extinction. Thus extinction occurs

Fig. 1.12—Superimposed isochromatic- and isoclinic-fringe patterns for a ring loaded in diametral compression.

when $n = \Delta/2\pi$, and by eq. (1.19) $n = N = (h/f_\sigma)(\sigma_1 - \sigma_2)$. When the principal-stress difference is such that $(h/f_\sigma)(\sigma_1 - \sigma_2) = 0, 1, 2$, etc., the conditions for extinction are met. The order of extinction ($N = 0, 1, 2$, etc.) more commonly referred to as the fringe order is dependent on the magnitude of the principal-stress difference, the thickness of the model, and the sensitivity of the photoelastic material.

In most practical cases the principal-stress difference $\sigma_1 - \sigma_2$ and the principal-stress directions change from one point to another in the model. If the analysis described here is extended to include every point in the model, two different lines of extinction will be formed. The first will be obtained when $\sigma_1 - \sigma_2 = Nf_\sigma/h$ with N equal to integers, and the second will occur when either the direction of σ_1 or the direction of σ_2 coincides with the axis of the polarizer. The first set of lines is known as isochromatic and the second set, isoclinic. These sets of lines are referred to as fringe patterns and are superimposed on each other as illustrated in Fig. 1.12.

Interpreting eq. (1.28) shows that the isoclinic and isochromatic fringes should be lines of zero width. However, inspection of the fringe patterns observed in the polariscope or in Fig. 1.12 shows the fringes as bands with a finite width. This fringe width is due to the recording characteristics of either the eye (in the case of direct observation) or of photographic film and paper. Exact minima in the intensity can be established within the region of a fringe by using a suitable photoelectric device to scan the width of a given fringe.

1.9 EFFECTS OF A STRESSED MODEL OR COATING IN A CIRCULAR POLARISCOPE

Dark Field

When a stressed photoelastic model is placed in the field of a circular polariscope in the manner shown in Fig. 1.13, the optical response is different from that observed in a plane polariscope. The circularly polarized light eliminates the isoclinic pattern and maintains the isochromatic pattern. Since the isochromatic pattern occurs without the isoclinics, it can be more easily observed and interpreted; for this reason, the circular polariscope is widely employed.

To show the effects of a stressed model in a circular polariscope, consider the light emerging from the series combination of a linear polarizer and a quarter-wave plate. When $\beta = \pi/4$, eq. (1.10) gives

$$A'_1 = (\sqrt{2}/2)\, k \sin(\omega t + \pi/2) = (\sqrt{2}/2)\, k \cos \omega t$$

$$A'_2 = (\sqrt{2}/2)\, k \sin \omega t \tag{1.29}$$

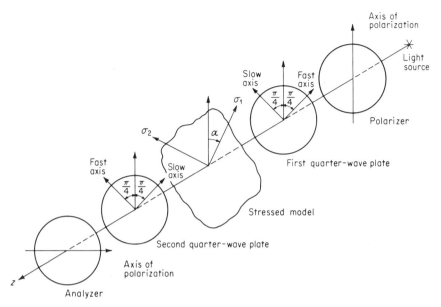

Fig. 1.13—Stressed photoelastic model in a circular polariscope (crossed polarizer and analyzer and crossed quarter-wave plate).

As this light vector emerges from the first quarter-wave plate, it is circularly polarized. The emerging light vector resulting from the combination of components A_1' and A_2' is of constant amplitude, and its tip sweeps out a circular helix as it propagates down the axis of the polariscope. In this analysis the individual components will be treated separately as they pass through the downstream elements in the polariscope.

The light components A_1' and A_2' from the first quarter-wave plate impinge on the model as shown in Fig. 1.14 and are resolved into components A_1'' and A_2'' along the σ_1 and σ_2 axes respectively. The magnitudes of A_1'' and A_2'' are given by

$$A_1'' = (\sqrt{2}/2)k \left[\cos \omega t \cos (\pi/4 - \alpha) + \sin \omega t \sin (\pi/4 - \alpha)\right]$$

$$= (\sqrt{2}/2)k \cos (\omega t + \alpha - \pi/4)$$

$$A_2'' = (\sqrt{2}/2)k \left[\sin \omega t \cos (\pi/4 - \alpha) - \cos \omega t \sin (\pi/4 - \alpha)\right]$$

$$= (\sqrt{2}/2)k \sin (\omega t + \alpha - \pi/4) \qquad (1.30)$$

Since the stressed model exhibits the optical characteristics of a retardation plate, the two components A_1'' and A_2'' propagate through the model with different velocities and emerge out of phase by an angle Δ, which is proportional to the stress difference as indicated in eq. (1.18).

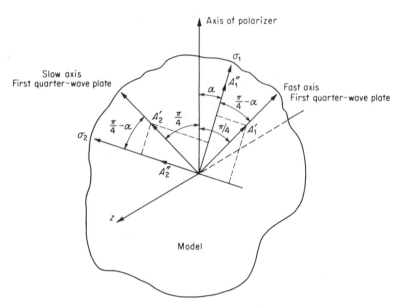

Fig. 1.14—Resolution of light components impinging on the stressed model.

Upon emergence the amplitudes become

$$A_1^{III} = (\sqrt{2}/2)k \cos (\omega t + \alpha - \pi/4 + \Delta)$$

$$A_2^{III} = (\sqrt{2}/2)k \sin (\omega t + \alpha - \pi/4) \qquad (1.31)$$

The light components A_1^{III} and A_2^{III} propagate from the model and impinge on the second quarter-wave plate (Fig. 1.15). The components resolved onto the fast and slow axes are denoted as A_1^{IV} and A_2^{IV} respectively, with amplitudes given by

$$A_1^{IV} = A_1^{III} \sin (\pi/4 - \alpha) + A_2^{III} \cos (\pi/4 - \alpha)$$

$$A_2^{IV} = A_1^{III} \cos (\pi/4 - \alpha) - A_2^{III} \sin (\pi/4 - \alpha) \qquad (1.32)$$

Assuming that the relative retardation of $\pi/2$ (which occurs as the light is transmitted through the quarter-wave plate) is applied in a positive sense to the A_1^{IV} component, the emerging components A_1^{V} and A_2^{V} are expressed as

$$A_1^{V} = (\sqrt{2}/2)k \left[\cos (\omega t + \alpha + \pi/4 + \Delta) \sin (\pi/4 - \alpha)\right.$$
$$\left. + \sin (\omega t + \alpha + \pi/4) \cos (\pi/4 - \alpha)\right]$$

$$A_2^{V} = A_2^{IV} = (\sqrt{2}/2) k \left[\cos (\omega t + \alpha - \pi/4 + \Delta) \cos (\pi/4 - \alpha)\right.$$
$$\left. - \sin (\omega t + \alpha - \pi/4) \sin (\pi/4 - \alpha)\right] \qquad (1.33)$$

Finally, the light enters the analyzer as illustrated in Fig. 1.16 and is resolved into vertical and horizontal components. The vertical

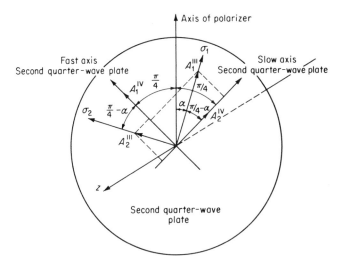

Fig. 1.15—Resolution of light components impinging on the second quarter-wave plate.

components are blocked by the analyzer, and the horizontal components are transmitted to give the output amplitude A where

$$A = (\sqrt{2}/2)\,(A_2^V - A_1^V) \tag{1.34}$$

Substituting eq. (1.33) into (1.34) and simplifying gives

$$A = \sqrt{2}k\,\cos(\omega t - 2\alpha)\sin(\Delta/2) \tag{1.35}$$

Square the amplitude to obtain the intensity of the emerging light as

$$I = K_1\,\cos^2(\omega t - 2\alpha)\sin^2(\Delta/2) \tag{1.36}$$

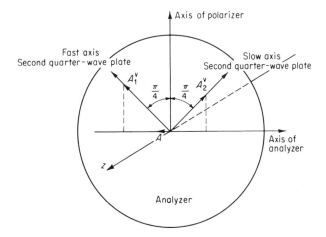

Fig. 1.16—Components of the light vector transmitted through the analyzer (dark field).

24   Chapter 1

Inspection of eq. (1.36) shows that the angle α, giving the direction of the principal stresses, occurs in the argument of the cosine term along with the ωt term; consequently, the isoclinic pattern does not develop. Indeed, the circular polariscope is employed solely to eliminate the isoclinic-fringe pattern from the superimposed isoclinic-isochromatic fringe pattern obtained in the plane polariscope.

The $\cos(\omega t - 2\alpha)$ term in eq. (1.36) has the same significance as discussed for the $\cos \omega t$ term in Section 1.8. The effect of the high-frequency periodic variation in the light intensity is averaged and, consequently, eq. (1.36) is rewritten as

$$I = K \sin^2 (\Delta/2) \tag{1.37}$$

From this result, extinction occurs when $\sin^2 (\Delta/2) = 0$; thus $I = 0$

Fig. 1.17—Dark-field isochromatic-fringe pattern for a ring in diametral compression.

only when $\Delta/2 = n\pi$ with $n = 0, 1, 2$, etc. This type of extinction is identical to that described in Section 1.8, Case III. The loci of these points of extinction produce an isochromatic pattern without the superimposed isoclinic pattern shown in Fig. 1.17. In this instance the circular polariscope is arranged with a dark-field setting, and the fringe order N is

$$N = \Delta/2\pi = n \qquad n = 0, 1, 2, \text{ etc.} \tag{1.38}$$

Thus the fringes are ordered with $N = n$ as $0, 1, 2$, etc.

Light Field

A circular polariscope is employed with both light and dark fields to provide additional data in a photoelastic analysis. The dark-field polariscope shown in Fig. 1.13 can be converted to a light field by rotating the analyzer through 90 deg. To determine the effect of a stressed model in a light-field circular polariscope, it is only necessary to consider the light components A_1^V and A_2^V as they enter the analyzer in its new orientation, as shown in Fig. 1.18.

The horizontal components of A_1^V and A_2^V will be absorbed, while the vertical components along the axis of polarization of the analyzer will be transmitted. The light vector emerging from the analyzer will lie in the vertical plane with an amplitude given by

$$A = (\sqrt{2}/2)(A_1^V + A_2^V) \tag{1.39}$$

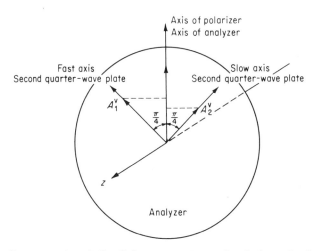

Fig. 1.18—Components of the light vector transmitted through the analyzer (light field).

Substituting eq. (1.33) into (1.39), expanding, and simplifying gives

$$A = \sqrt{2}k \cos \omega t \cos (\Delta/2)$$

and an intensity

$$I = K_1 \cos^2 \omega t \cos^2 (\Delta/2) \tag{1.40}$$

Again, the term $\cos \omega t$ is treated by a time average over the observation period so that eq. (1.40) is rewritten as

$$I = K \cos^2 (\Delta/2) \tag{1.41}$$

This result indicates that extinction ($I = 0$) will occur when

$$\Delta/2 = (1 + 2n) (\pi/2) \qquad n = 0, 1, 2, \text{etc.}$$

Fig. 1.19—Light-field isochromatic-fringe pattern for a ring in diametral compression.

Noting that

$$N = \Delta/2\pi = (1 + 2\,n)/2 \qquad (1.42)$$

it is evident that the fringe order N increases as 1/2, 3/2, 5/2, etc., as n increases as 0, 1, 2, etc. An isochromatic-fringe pattern for a light-field polariscope is illustrated in Fig. 1.19.

The circular polariscope used in the light and dark fields provides two independent sets of isochromatic fringe patterns that give entire field representation of the order of fringes to the nearest 1/2 order. While the analysis presented in these sections has been for a transmission-type polariscope, the development is the same for reflection-type polariscopes used with photoelastic coatings.

1.10 WHITE LIGHT

It was shown above that conditions for extinction are met in plane and circular polariscopes when

$$\sin^2 (\Delta/2) = 0 \qquad \text{or} \qquad \Delta/2 = N\pi$$

where N is an integer also called fringe order. At all such points the retardation δ is

$$\delta = h\,C\,(\sigma_1 - \sigma_2) = (\lambda/2\pi)\Delta = N\lambda \qquad \text{or} \qquad \delta = 0, 1\lambda, 2\lambda, \text{etc.} \qquad (1.43)$$

When a white-light source is used in a polariscope, many different frequencies and wavelengths of light are simultaneously present. At a point eq. (1.43) cannot be satisfied for all different wavelengths. When $\delta = 0$, extinction will occur for all wavelengths and a black isochromatic fringe will be observed. For $\delta \neq 0$ (e.g., $\delta = 5200$ Å) the extinction will occur only for the green of wavelength $\lambda = 5200$ Å. A mixture of other wavelengths produced by the source will be observed, producing a visual observation of red. The scale of observed colors as the retardation increases is shown in Table 2.2 and Fig. 2.4.

Since only points where $\delta = 0$ appear black (complete extinction) in white light, two important observations should be made:

1. In the plane polariscope only isoclinic lines (σ_1 or σ_2 parallel to the polarizer) or $N = 0$ lines ($\sigma_1 - \sigma_2 = 0$) appear black, simplifying recognition and photography of isoclinic lines.

2. In the circular dark-field polariscope only points where $N = 0$ and $\delta = 0$ ($\sigma_1 - \sigma_2 = 0$) points will produce a black fringe. This property is used to recognize and measure fringe orders when using compensators, as described in Chapters 2 and 4.

1.11 TARDY METHOD OF MEASURING FRACTIONAL
FRINGE ORDERS

The Tardy method of compensation is very commonly employed to determine the fractional order of the fringe at any point on the model. To employ the Tardy method, the model (or coating) is aligned in the circular polariscope so that the principal-stress directions coincide with the axes of the polarizer and analyzer. With this alignment, $\alpha = 0$ as indicated in Fig. 1.14, and eqs. (1.33) and (1.34) simplify to yield

$$A_1^V = (k/2) [\cos (\omega t + \pi/4 + \Delta) + \sin (\omega t + \pi/4)]$$

$$A_2^V = (k/2) [\cos (\omega t - \pi/4 + \Delta) - \sin (\omega t - \pi/4)]$$

Using trigonometric identities, the above expression can be simplified further:

$$A_1^V = k \cos (\omega t + \Delta/2) \cos (\Delta/2 + \pi/4)$$

$$A_2^V = k \cos (\omega t + \Delta/2) \sin (\Delta/2 + \pi/4) \qquad (1.44)$$

Since there is no phase shift between components A_1^V and A_2^V, the light that emerges from the second quarter-wave plate is plane polarized. The angle that the light vector A makes with the quarter-wave plate axis is $\Delta/2 + \pi/4$ as shown on Fig. 1.20, and its angle to the polarizer axis is $\Delta/2$.

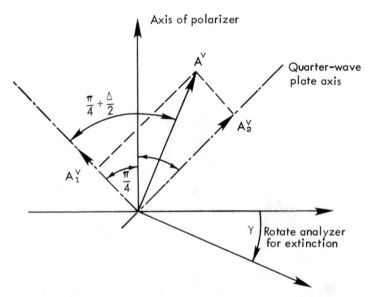

Fig. 1.20—Rotation of the analyzer to obtain extinction in the Tardy method of compensation.

Extinction can be obtained by rotating the analyzer until it is perpendicular to the light vector A. From Fig. 1.20 it can be seen that the rotation γ required is

$$\gamma = \Delta/2 \tag{1.45}$$

The fringe order N will consist in general of an integer n and a fraction r.

$$N = n + r = \Delta/2\pi \quad \text{or} \quad \Delta = 2\pi\,(n + r)$$

Substituting for Δ in eq. (1.45) yields the expression relating the angle γ to the fringe order:

$$\gamma = \pi\,(n + r) = \pi n + \pi r \tag{1.46}$$

In practice the rotation of the analyzer is made to achieve extinction, and the angle γ is measured. The fractional fringe order r is then related to γ by $r = \gamma/\pi$. Further details of the Tardy method are discussed in Chapter 2.

1.12 SUMMARY

This chapter has introduced the properties of light necessary to explain the photoelastic effect and to derive the equations associated with the elementary theory of elasticity. The key equations derived are summarized below:

The relative retardation

$$\delta = h\,(n_1 - n_2)$$

or

$$\Delta = (2\pi h/\lambda)\,(n_1 - n_2)$$

The stress-optic law, which may be expressed as

$$\sigma_1 - \sigma_2 = (\Delta/2\pi)\,(\lambda/c)\,(1/h)$$

or

$$\sigma_1 - \sigma_2 = Nf_\sigma/h$$

The relations between the stresses and strains in the coating and the optical response

$$\sigma_1^c - \sigma_2^c = Nf_\sigma/2h_c = NF_\sigma$$

and

$$\epsilon_1^c - \epsilon_2^c = (1 + \nu_c)\,f_\sigma N/2Eh_c = f_\epsilon\,N/2h = F_\epsilon N$$

The intensity of light emerging from a plane polariscope is

$$I = K_1 \sin^2 2\alpha \sin^2 (\Delta/2) \cos^2 \omega t$$

and from a circular polariscope with dark-field settings

$$I = K_1 \cos^2 (\omega t - 2\alpha) \sin^2 (\Delta/2)$$

and with light-field settings

$$I = K_1 \cos^2 \omega t \cos^2 (\Delta/2)$$

Finally the fractional fringe order measured by the Tardy method is $r = \gamma/\pi$.

2
ELEMENTARY THEORY OF
PHOTOELASTIC COATINGS

When a photoelastic coating is bonded to the surface of a specimen, the surface displacements are transmitted to the coating through the adhesive. The exact distribution of the stresses and strains through the thickness of the coating is a complex determination; it is discussed fully in Chapter 5.

Consider that both the coating and the specimen are subjected to a state of plane stress where the strains are uniformly distributed through the thickness of the coating. With this simpler plane-stress approach, it is possible to introduce the relations between stresses in the coating and the specimen, the optical response of the coating, and the coating sensitivity. Methods for measuring the fringe orders and separating the principal stresses are then described.

2.1 STRESSES AND STRAINS IN THE COATING[10–12]

Consider a specimen with a thin coating, as illustrated in Fig. 2.1, with a two-dimensional geometry and a loading that produces a state of plane stress. The stress and strain fields in the specimen are independent of the position parameter Z, and

$$\sigma^s_{xx} = \sigma^s_{xx}(x, y) \qquad \epsilon^s_{xx} = \epsilon^s_{xx}(x, y)$$

$$\sigma^s_{yy} = \sigma^s_{yy}(x, y) \qquad \epsilon^s_{yy} = \epsilon^s_{yy}(x, y)$$

$$\tau^s_{xy} = \tau^s_{xy}(x, y) \qquad \epsilon^s_{zz} = \epsilon^s_{zz}(x, y)$$

$$\tau^s_{xz} = \tau^s_{yz} = \sigma^s_{zz} = 0 \qquad \gamma^s_{xy} = \gamma^s_{xy}(x, y), \gamma^s_{xz} = \gamma^s_{yz} = 0 \qquad (2.1)$$

Because the strains in the specimen are constant with respect to z, it is assumed that they are transmitted to the coating through the adhesive without loss or amplification and that they are also constant

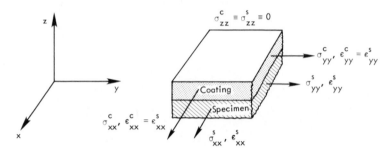

Fig. 2.1—Plane specimen with a thin photoelastic coating.

with respect to z through the thickness of the coating. Thus

$$\epsilon_{xx}^c (x, y) = \epsilon_{xx}^s (x, y) \qquad \epsilon_{yy}^c (x, y) = \epsilon_{yy}^s (x, y) \tag{2.2}$$

where superscripts c and s refer to the coating and the specimen respectively.

Because the coating is relatively thin, it is assumed that a state of plane stress also exists in the coating, from which

$$\sigma_{zz}^c = \sigma_{zz}^s = 0 \tag{2.3}$$

The foregoing assumptions neglect the influence of (1) the coating reinforcement on the specimen behavior, (2) any mismatch in Poisson's ratio between the specimen and coating, and (3) the strain ϵ_{zz}^s normal to the plane of the specimen. These influences, which are usually small, are treated in Chapter 5.

The stresses in the coating can be expressed as functions of the stresses in the specimen by combining eqs. (2.2) with the stress-strain relationships as follows:

$$\epsilon_1^s = (1/E_s) (\sigma_1^s - \nu_s \sigma_2^s) = \epsilon_1^c = (1/E_c) (\sigma_1^c - \nu_c \sigma_2^c)$$

$$\epsilon_2^s = (1/E_s) (\sigma_2^s - \nu_s \sigma_1^s) = \epsilon_2^c = (1/E_c) (\sigma_2^c - \nu_c \sigma_1^c) \tag{2.4}$$

where subscripts 1 and 2 refer to the principal directions. Rewriting eqs. (2.4),

$$\sigma_1^c = \frac{E_c}{E_s(1 - \nu_c^2)} [(1 - \nu_c \nu_s)\sigma_1^s + (\nu_c - \nu_s)\sigma_2^s] \tag{2.5a}$$

$$\sigma_2^c = \frac{E_c}{E_s(1 - \nu_c^2)} [(1 - \nu_c \nu_s)\sigma_2^s + (\nu_c - \nu_s)\sigma_1^s] \tag{2.5b}$$

$$\sigma_1^c - \sigma_2^c = (E_c/E_s) [(1 + \nu_s)/(1 + \nu_c)] (\sigma_1^s - \sigma_2^s) \tag{2.5c}$$

$$\epsilon_1^c - \epsilon_2^c = \epsilon_1^s - \epsilon_2^s \tag{2.5d}$$

Thus the stresses in the coating are linearly related to the stresses in the specimen, with the elastic constants of both materials entering into the proportionality coefficients. In addition, the principal-stress and principal-strain directions in the coating are colinear with those in the specimen.

2.2 OPTICAL RESPONSE OF THE COATING[4,13]

When a coated specimen under load is examined using a *plane reflection polariscope*, the state of stress in the photoelastic coating will produce two superimposed fringe patterns. One group, called *isoclinic fringes*, is indicative of the principal-stress directions. These fringes represent lines along which the directions of the principal stresses σ_1 or σ_2 have a constant angular inclination to an arbitrary reference axis, as illustrated in Fig. 2.2. A black isoclinic fringe is generated wherever the directions of the principal stresses coincide with the axis of polarization of either the polarizer or analyzer. By rotating the polarizer and analyzer together while maintaining them crossed, all possible families of isoclinic fringes can be generated. This procedure permits determining the principal-stress directions over the entire field of the coating. Because the directions of the principal stresses in the coating and specimen coincide, the isoclinic data from the coating give the principal-stress directions in the specimen directly.

The second family of fringes observed with a plane polariscope consists of *isochromatic fringes*. These represent lines or contours along which the difference in principal stresses in the coating

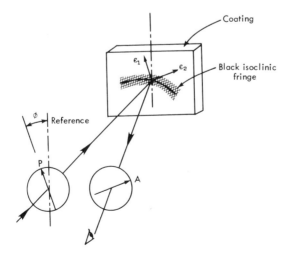

Fig. 2.2—Formation of isoclinic fringes in a photoelastic coating.

$(\sigma_1^c - \sigma_2^c)$ is constant in magnitude. When white light is employed with the plane reflection polariscope, the isochromatic fringes are brightly colored bands (except for the zero-order fringe, which is black like the isoclinics). The colors characteristic of the isochromatic pattern and the fact that this pattern remains stationary when the polarizer and analyzer are rotated together, permit the observer to distinguish readily between the isoclinic and isochromatic fringes. A *circular reflection polariscope*, which incorporates two quarter-wave retardation plates in addition to the polarizer and analyzer, is normally used to obtain data from the isochromatic-fringe pattern. The effect of the quarter-wave plates is to induce circular polarization, thus eliminating the isoclinic-fringe pattern without disturbing the isochromatic pattern.

The isochromatic-fringe pattern can be looked upon as a full-field "topographical" map of the stresses, in which the individual contour lines are identified by the fringe-order number N. The difference in the principal stresses at each point in the coating is related to the fringe order N by the classical *stress-optic law*—modified to account for double passage of the light through the coating. Thus

$$\sigma_1^c - \sigma_2^c = N f_\sigma / 2 h_c \qquad (2.6)$$

where f_σ = material-fringe value (stress), psi-in./fringe (MPa-mm/
            fringe)
        h_c = coating thickness, in. (mm)

The difference in principal stresses at a point in the specimen can be related to the fringe order by combining eqs. (2.5b) and (2.6) to give

$$\sigma_1^s - \sigma_2^s = (E_s/E_c) \left[(1 + \nu_c)/(1 + \nu_s) \right] (N f_\sigma / 2 h_c) \qquad (2.7)$$

Similarly, employing the strain-optic law and the identity of the specimen and coating strains as expressed by eq. (2.5d), a relationship can be obtained between the difference in principal strains and the fringe order:

$$\epsilon_1^s - \epsilon_2^s = \epsilon_1^c - \epsilon_2^c = N f_\epsilon / 2 h_c \qquad (2.8)$$

where f_ϵ is the material-fringe value (strain), in./fringe (mm/fringe).

The strain-optic relationship presented in eq. (2.8) is often expressed in a slightly different form, with the constant f_ϵ replaced by λ/K to give

$$\epsilon_1^s - \epsilon_2^s = \epsilon_1^c - \epsilon_2^c = (N/2h_c) (\lambda/K) \qquad (2.9)$$

where λ = wavelength of light with which the coating is illuminated
        K = optical-sensitivity constant (dimensionless)

For a perfectly elastic photoelastic material, f_ϵ, f_σ, and K are related as follows:

$$f_\epsilon = \left(\frac{1 + \nu_c}{E_c}\right) f_\sigma \qquad f_\epsilon = \frac{\lambda}{K} \qquad f_\sigma = \left(\frac{E_c}{1 + \nu_c}\right)\left(\frac{\lambda}{K}\right) \qquad (2.10)$$

The alternative form of the strain-optic law [eq. (2.9)], in which the coating sensitivity is expressed by K, is more general since it can be employed with different light sources for which λ may vary.

The isochromatic and isoclinic data are employed together to establish the shear strain γ^c_{xy}. Using the identity between coating and specimen shear strains and introducing the strain transformation relationship,

$$\gamma^c_{xy} = \gamma^s_{xy} = (\epsilon^c_1 - \epsilon^c_2) \sin 2\phi \qquad (2.11)$$

where ϕ is the isoclinic parameter defining the angle between σ_1 and the x axis.

From eqs. (2.8) and (2.10),

$$\gamma^s_{xy} = (Nf_\epsilon/2h_c) \sin 2\phi \qquad (2.12)$$

and from the shear-stress/shear-strain relationship $\tau^s_{xy} = G_s \gamma^s_{xy}$, the shear stress in the specimen is given by:

$$\tau^s_{xy} = \frac{G_s Nf_\epsilon}{2h_c} \sin 2\phi = \frac{E_s}{E_c} \frac{1 + \nu_c}{1 + \nu_s} \frac{Nf_\sigma}{4h_c} \sin 2\phi \qquad (2.13)$$

The photoelastic data obtained from the isochromatic and isoclinic patterns do not provide sufficient information to determine the individual magnitudes of the principal stresses or strains. Auxiliary methods—generally required for determining σ_1, σ_2, and τ_{max}—are described in Section 2.5.

2.3 COATING SENSITIVITY[14]

The optical response of a photoelastic coating to the stress field in a coated specimen can be evaluated by defining a *stress-sensitivity index* S_σ as follows:

$$S_\sigma = N/(\sigma^s_1 - \sigma^s_2) \qquad (2.14)$$

By substituting eqs. (2.7) and (2.11) into eq. (2.14), the stress-sensitivity index can be written as

$$S_\sigma = (2h_c/f_\epsilon)\,[(1 + \nu_s)/E_s] = C_c C_s \qquad (2.15)$$

where $C_c = 2h_c/f_\epsilon$, the coating coefficient of sensitivity
    $C_s = (1 + \nu_s)/E_s$, the specimen coefficient of sensitivity

From eq. (2.15) for the stress-sensitivity index, it is evident that the optical response depends on coating parameters embodied in C_c and on the elastic constants of the specimen material as reflected in C_s. The value of C_c can be adjusted by changing the coating thickness, but arbitrary changes in h_c are not usually practical because of errors that may be introduced by excessively thick coatings.

In application of the theory of photoelastic coatings to problems in elastic-stress analysis, the maximum achievable response of the coating is necessarily limited to that occurring when some point on the specimen first reaches the yield condition. If $\sigma_2^s < 0$ and the material follows the Tresca yield criterion, the maximum difference in principal stresses at yielding is

$$(\sigma_1^s - \sigma_2^s)_{\max} = S_{yp} \qquad (2.16)$$

where S_{yp} is the specimen material yield strength. The corresponding maximum fringe order exhibited by the coating (at incipient yielding) follows from eqs. (2.14), (2.15), and (2.16):

$$N_{\max} = C_c C_s S_{yp} = (2h_c/f_\epsilon)\,[(1 + \nu_s)/E_s]\,S_{yp} \qquad (2.17)$$

In elastic stress analysis it can be seen from eq. (2.17) that the maximum achievable fringe order in the photoelastic coating is a function of three parameters—S_{yp}, C_s, and C_c. Typical values of the maximum achievable fringe order are given in Table 2.1 for a wide range of materials. In computing these values of $N_{\max}$, it was assumed that $h_c = 0.1$ in. (2.54 mm) and $K = 0.14$—the latter corresponding to $f_\epsilon = 1.62 \times 10^{-4}$ in./fringe (41.1×10^{-4} mm/fringe) for a wavelength of light equal to 22.7×10^{-6} in. (577 nm).

From the foregoing and Table 2.1, it can be seen that the optical response of the coating to specimen stresses is highly dependent upon the mechanical properties of the specimen material. The maximum achievable fringe orders for elastic-stress analysis range over two decades of magnitude—from very low response for coatings on glass or concrete to extremely high response on high-strength steel, aluminum, and certain plastics such as nylon and fiberglass. In practice, the actual coating thickness used and the method employed to determine the fringe order will depend upon the nature of the specific problem. Where the optical response is low, thicker coatings will generally be required, and it will usually be necessary to employ compensation for determining the fringe order. With higher strength substrate materials thin coatings can be employed, and fringe orders can be observed directly on the model or in a photograph of the fringe pattern.

Table 2.1—Typical Values of the Maximum Fringe Order That
Can Be Achieved in an Elastic-Stress Analysis

Material	S_{yp} MPa	(psi)	C_s $(1/psi \times 10^7)$	N^*_{max} (Fringes)
Steel—H.R. 1020	240	35,000	0.43	1.86
C.D. 1020	310	45,000	0.43	2.40
H.T. 1040	550	80,000	0.43	4.26
H.T. 4140	900	130,000	0.43	6.92
H.T. 52100	1250	180,000	0.43	9.58
Maraging (18-Ni)	1720	250,000	0.49	15.2
Aluminum—1100 H16	130	20,000	1.33	3.30
3004 H34	200	29,000	1.33	4.78
2024 T3	350	50,000	1.33	8.25
7075 T6	500	73,000	1.33	12.0
Magnesium—AM 11	145	21,000	2.08	5.42
Cartridge brass	430	63,000	0.87	6.80
Phosphor bronze	520	75,000	0.87	8.10
Beryllium copper	480	70,000	0.77	6.70
Glass	21	3,000	1.25	0.46
Concrete (compression)	28	4,000	4.20	2.10
Plastic nylon 6-6	80	12,000	46.7	69.0
Glass-reinforced plastic (isotropic)	800	120,000	3.18	47.4

*For $C_c = 1.24 \times 10^3$ fringe.

2.4 FRINGE-ORDER DETERMINATION

The method utilized in any particular case for determining the
fringe orders in the coating depends upon the response of the coating
and the accuracy (or resolution) required in the analysis. A detailed
discussion of photoelastic instruments for fringe order determination
is presented in Section 4.3.

If the response of the coating is large (e.g., four or more fringes),
monochromatic light can be used to obtain photographs of the light-
and dark-field isochromatic-fringe patterns. Fringe orders can usually be interpolated or extrapolated to the nearest 0.2 fringe. If the
maximum fringe order is four, this means that a resolution of 5 percent is possible. An example of a typical high-response fringe pattern
that can be analyzed in this manner is shown in Fig. 2.3.

For fringe patterns with two to four fringes it is normally advantageous to use white light, producing colored patterns. The colored
pattern is caused by the attenuation and extinction of one or more
colors from the white-light spectrum, and the observed colored
fringes represent the complementary colors produced by the transmitted portion of the spectrum. Plate C.1 illustrates the sequence of

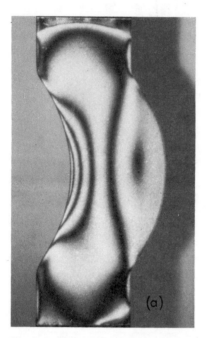 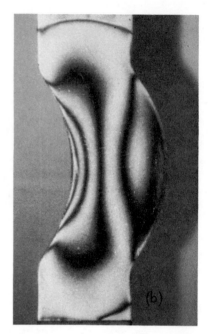

Fig. 2.3—High-order fringe pattern permits direct full-field analysis from the photographs: (a) dark field, and (b) light field.

colored fringes produced in a field with a linearly increasing stress magnitude. Although the exact color "shade" associated with each stress level is a function of the energy distribution in the white-light spectrum and the recording characteristics of the color film used, the color sequence listed in Table 2.2 is an adequate guide for most practical purposes. Note that the tint-of-passage fringes occur at multiples of 5750 Å (22.7×10^{-6} in.) retardation; the wavelength 5750 Å is therefore considered effective when white light is used, and the retardation at tint-of-passage 1, 2, 3, etc., is λ, 2λ, 3λ, etc.

As demonstrated by Table 2.2, the use of white light substantially increases the number of fringes that can be identified. For example, in the interval from $0 \leqslant N \leqslant 2$, there are 12 distinct color bands that can be used to establish fractional fringe orders. Moreover, the polariscope can be used in the light-field mode of operation to yield a second family of colored fringes, effectively doubling the amount of data available for estimating fractional fringe orders. By these procedures the fringe orders can be established to within about 0.1 fringe, giving a resolution of approximately 5 percent for a total of two fringes.

For precise fringe-order determinations where the highest fringe order is less than two and a resolution of 5 percent or better is re-

Table 2.2—Sequence of Colors Produced in a Dark-field White-light Polariscope

Color	Retardation (Å)	Fringe Order
Black	0	0
Gray	1,600	0.28
White	2,600	0.45
Yellow	3,500	0.60
Orange	4,600	0.79
Red	5,200	0.90
Tint of passage 1*	5,770	1.00
Blue	6,200	1.06
Blue-green	7,000	1.20
Green-yellow	8,000	1.38
Orange	9,400	1.62
Red	10,500	1.81
Tint of passage 2*	11,500	2.00
Green	13,500	2.33
Green-yellow	14,500	2.50
Pink	15,500	2.67
Tint of passage 3*	17,300	3.00
Green	18,000	3.10
Pink	21,000	3.60
Tint of passage 4*	23,000	4.00
Green	24,000	4.13

*The tint of passage is a sharp dividing zone occurring between red and blue in the first-order fringe, red and green in the second-order fringe, and pink and green in the third-, fourth-, and fifth-order fringes. Beyond five fringes, white-light analysis is not adequate.

quired in the analysis, compensation techniques are necessary. Although compensation methods greatly improve the resolution in determining fringe orders, they are point-by-point methods and require additional operations by the investigator.

The principle of compensation is illustrated in Fig. 2.4 where three primary elements are shown. Element A represents a general state of stress in the coating that would produce some unknown fringe order N. Element B is aligned with element A and subjected to a uniaxial state of stress ($\sigma_0 = \sigma_1^c - \sigma_2^c$) that is coaxial with σ_2 on element A. By superimposing the stresses on elements A and B, the effect is to produce an isotropic state of stress as shown in element C, where both the principal stresses are σ_1^c. The superposition of σ_0 onto σ_2^c gave $\sigma_2^c + (\sigma_1^c - \sigma_2^c) = \sigma_1^c$, thus providing the isotropic state with $N = 0$. By measuring the stress σ_0 necessary to produce the photoelastic null state, a precise determination of $\sigma_1^c - \sigma_2^c$ is made.

In practice, compensation is accomplished by replacing element B by an optical device that produces the same optical effect as the

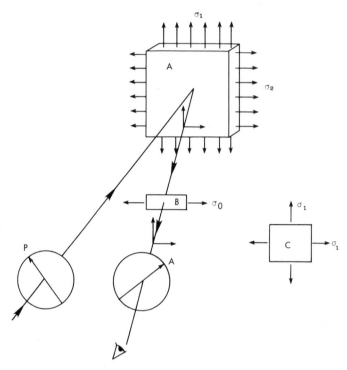

Fig. 2.4—Superposition of stress states in elements A and B to produce a photo-elastic null state in element C.

stressed element. The *Babinet-Soleil* compensator is a very effective optical accessory to the polariscope for precise determination of fringe orders. This instrument contains a quartz plate of uniform thickness t_1 and two quartz wedges, as illustrated in Fig. 2.5. The optic axes of the plate and the quartz wedges are mutually orthogonal. Because quartz is a permanently doubly refractive material, the relative retardation produced by the compensator can be precisely controlled by adjusting the total thickness t_2 of the two wedges with a calibrated micrometer screw. When $t_1 = t_2$, the net relative retardation is zero in the compensator. However, for $t_2 \gtrless t_1$, either positive or negative retardation can be superimposed on the polariscope system.

To employ the Babinet-Soleil compensator, a point on the coated specimen or test object is selected and the isoclinic parameter is measured there to determine the principal-stress directions. The compensator is then introduced into the polariscope and aligned with one of the principal-stress directions. By adjusting the micrometer screw on the compensator, the strain-induced retardation in the photoelas-

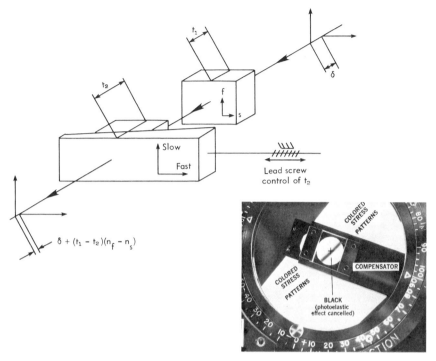

Fig. 2.5—Schematic diagram of a Babinet-Soleil compensator.

tic coating can be canceled or nullified through placing a controlled (and known) retardation of the opposite sign in series with it. The reading of the micrometer is proportional to the fringe order at the point under study, and the direction of screw rotation necessary to establish the null indicates whether the compensator is aligned along the σ_1 or σ_2 direction. Determination of N to within 0.01 fringe can be achieved with the Babinet-Soleil compensator, resulting in a resolution of about 2 percent when the maximum fringe order in the coating is 0.5.

Compensation can also be effected optically without auxiliary equipment by employing the analyzer of the polariscope to serve as the compensating device. This technique is known as the *Tardy* method. The first step in the procedure is to align the polarizer with the direction of one of the principal stresses σ_1 or σ_2 at the point of interest. All other elements of the polariscope are then rotated relative to the polarizer so that a standard dark-field circular polariscope exists. The components of the light rays emerging from the second quarter-wave plate are canceled by rotating the analyzer through an angle γ as indicated in Fig. 2.6. The fringe order at the point under

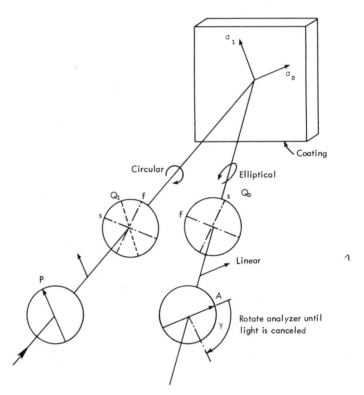

Fig. 2.6—Analyzer rotation in the Tardy method to adjust γ for extinction.

study is given by

$$N = n \pm \gamma/\pi \tag{2.18}$$

To use eq. (2.18), the value of n must be determined separately and the $\pm$ choice decided from the motion of the dark-field isochromatic pattern with respect to the point of interest. For example, in Fig. 2.7 point Q_1 is located between fringes of order 1 and 2. In performing compensation at point Q_1, n is taken as one and the analyzer is rotated through the angle γ to move the first-order fringe over Q_1. The actual fringe order at Q_1 then is $N = 1 + \gamma/\pi$. Alternatively, compensation can be effected at point Q_1 by rotating the analyzer in the opposite direction through the angle γ_1 to move the second-order fringe to Q_1. In this case the fringe order at Q_1 is $N = 2 - \gamma_1/\pi$. From the direction of rotation required to move the lower (or higher) order over the point, the orientation of σ_1 can be determined.

The fringe orders at all points located on a particular isoclinic can be quickly determined with this method by simply rotating the ana-

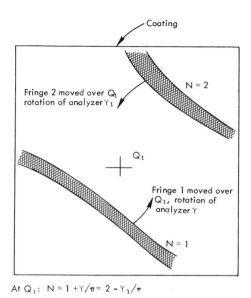

At Q_1: $N = 1 + \gamma/\pi = 2 - \gamma_1/\pi$

Fig. 2.7—Fractional fringe-order determination at point Q by the Tardy compensation method.

lyzer as necessary to bring the nearest fringe over each point of interest. Similarly, the fringe orders at points $R_i, \ldots, R_n$ along a different isoclinic can be readily established after the polariscope has been realigned in the ϕ_2 direction.

Fringe orders can be measured by the Tardy method to within about 0.02 fringe, yielding a resolution of about 4 percent for a maximum fringe order of 0.5 in the coating. (Similar results can be achieved with the Senarmont method, which is described in most standard reference texts on photoelasticity.) The Babinet-Soleil method is preferred over the Tardy method because it is more sensitive and does not require the identification of the nearest integral fringe order.

2.5 STRESS-SEPARATION METHODS

Examination of eqs. (2.7) and (2.8) shows that the isochromatic data permit the determination of either $\sigma_1^s - \sigma_2^s$ or $\epsilon_1^s - \epsilon_2^s$, but not the individual principal stresses or strains. In case of uniaxial stress where one of principal stresses is zero, eqs. (2.7) and (2.8) provide the complete solution. This will occur on all free boundaries and on discontinuities where most critical measurements are required and highest stresses usually occur. Also, in many instances one of the stresses is small or negligible. There again the measurement of the difference of principal stresses provides the "stress" with sufficient engineering approximations. However, in many applications additional information is needed to "separate" (obtain the individual values of) σ_1^s and σ_2^s or ϵ_1^s and ϵ_2^s. Among the several methods developed for separating the principal stresses or strains are (1) oblique incidence, (2) strip coatings, and (3) graphical integration. Strain gages can also be used very effectively for the same purpose.

Oblique Incidence[15]

In the derivation of eqs. (2.7) and (2.8) it was implicitly assumed that the incident and reflected light rays in the polariscope were normal to the surface of the coated specimen, or very nearly so. If a second measurement is made with the incident and reflected rays oblique to the surface, the second set of data required for separating the principal stresses and strains can be obtained. The relationships for determining individual principal stresses and strains from combined normal- and oblique-incidence fringe-order measurements are derived here.

Rewriting eq. (2.8), a normal-incidence fringe-order determination gives

$$N_0 = (2h_c/f_\epsilon)\,(\epsilon_x - \epsilon_y)$$

where superscripts on the strain terms have been omitted since it is assumed in this derivation that the coating and specimen strains are identical. The x and y are directions of principal strains as established from isoclinic determination. If an oblique-incidence measurement N_{θ_x} is made by rotating the direction of observation about σ_x by the angle θ, as illustrated in Fig. 2.8, the fringe order is expressed by

$$N_{\theta_x} = (2h_c/f_\epsilon \cos \theta)\,(\epsilon_x - \epsilon_y^*) \tag{2.19}$$

where ϵ_y^* is the secondary principal strain in the planes normal to the (oblique) incident and reflected light rays.

From the relationships for strain transformation at a point, the secondary strain ϵ_y^* is given by:

$$\epsilon_y^* = \epsilon_y \cos^2 \theta + \epsilon_z \sin^2 \theta \tag{2.20}$$

And since the coating is under a plane state of stress,

$$\epsilon_z = -\,[\nu_c/(1 - \nu_c)]\,(\epsilon_x + \epsilon_y) \tag{2.21}$$

Substituting eqs. (2.20) and (2.21) into eq. (2.19) yields

$$N_{\theta_x} = \frac{2h_c}{f_\epsilon}\,\frac{1}{(1 - \nu_c)\cos \theta}\,[\epsilon_x(1 - \nu_c \cos^2 \theta) - \epsilon_y(\cos^2 \theta - \nu_c)] \tag{2.22}$$

If the rotation of the direction of observation is made around σ_y a

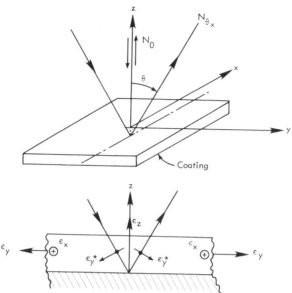

Fig. 2.8—Coordinates for oblique-incidence measurements.

similar expression can be derived:

$$N_{\theta_y} = \frac{2h_c}{f_\epsilon} \frac{1}{(1 - \nu_c) \cos \theta} [\epsilon_y (1 - \nu_c \cos^2 \theta) - \epsilon_x (\cos^2 \theta - \nu_c)] \quad (2.23)$$

Simultaneous solution of eqs. (2.8) and (2.22) for ϵ_x and ϵ_y provides the necessary relationships for determining the individual principal strains from the normal- and oblique-incidence fringe-order measurements:

$$\epsilon_x = \frac{f_\epsilon}{2h_c} \frac{1}{(1 + \nu_c) \sin^2 \theta} [N_{\theta_x}(1 - \nu_c) \cos \theta - N_0(\cos^2 \theta - \nu_c)]$$

$$\epsilon_y = \frac{f_\epsilon}{2h_c} \frac{1}{(1 + \nu_c) \sin^2 \theta} [N_{\theta_x}(1 - \nu_c) \cos \theta - N_0(1 - \nu_c \cos^2 \theta)]$$

$$(2.24)$$

An alternate procedure consists of obtaining two measurements in oblique incidence, rotating the direction of observation about σ_x by the angle θ and then rotating about σ_y by the same angle. Combining eqs. (2.22) and (2.23) yields

$$\epsilon_x = \frac{f_\epsilon}{2h_c} \frac{\cos \theta}{(1 + \nu_c)(1 - \cos^4 \theta)} [N_{\theta_x}(1 - \nu_c \cos^2 \theta) + N_{\theta_y}(\cos^2 \theta - \nu_c)]$$

$$\epsilon_y = \frac{f_\epsilon}{2h_c} \frac{\cos \theta}{(1 + \nu_c)(1 - \cos^4 \theta)} [N_{\theta_x}(\cos^2 \theta - \nu_c) + N_{\theta_y}(1 - \nu_c \cos^2 \theta)]$$

$$(2.25)$$

The derivation of eqs. (2.24) and (2.25) implies that the measured fringe orders N_0, N_{θ_x} are positive ($N > 0$) when $\epsilon_x - \epsilon_y$ or $\epsilon_x - \epsilon_y^*$ are positive. The sign of N can be quickly determined by using a compensator. Once the equations are solved, the algebraically larger strain ϵ_x or ϵ_y becomes the maximum principal strain ϵ_1.

Equations (2.24) can be rewritten into a general form:

$$\epsilon_x = (f_\epsilon/2h_c)\,(A N_{\theta_x} - B N_0)$$

$$\epsilon_y = (f_\epsilon/2h_c)\,(A N_{\theta_x} - C N_0)$$

with the coefficients A, B, and C being functions of Poisson's ratio of the coating material and of the angle θ. Since the physical properties of the coating material are not always known, it is customary to establish A, B, and C by calibration, measuring N_0, N_{θ_x}, and N_{θ_y} on a specimen where ϵ_x and ϵ_y are known. Solving A, B, and C in terms of ϵ_x, ϵ_y and N_0, N_θ yields

$$A = [(\epsilon_x + \epsilon_y)/(\epsilon_x - \epsilon_y)]\ [N_0/(N_{\theta_x} + N_{\theta_y})]$$

$$B = [(\epsilon_x + \epsilon_y)/(\epsilon_x - \epsilon_y)]\ [(N_{\theta_x} - N_{\theta_y})/(N_{\theta_x} + N_{\theta_y})] - (1/2)$$

$$C = B + 1 \tag{2.26}$$

Substitution of eq. (2.24) into the biaxial stress-strain relationships,

$$\sigma_1^s = \frac{E_s}{1 - \nu_s^2}\,(\epsilon_1 + \nu_s\epsilon_2) \qquad \sigma_2^s = \frac{E_s}{1 - \nu_s^2}\,(\epsilon_2 + \nu_s\epsilon_1) \tag{2.27}$$

yields expressions for the individual principal stresses in the specimen:

$$\sigma_1^s = \frac{E_s f_\epsilon}{2h_c(1 + \nu_c)\sin^2\theta}\,(F_3 \cos\theta\, N_{\theta 1} - F_2 \cos^2\theta\, N_0)$$

$$\sigma_2^s = \frac{E_s f_\epsilon}{2h_c(1 + \nu_c)\sin^2\theta}\,(F_3 \cos\theta\, N_{\theta 1} - F_1\, N_0) \tag{2.28}$$

where

$$F_1 = [1/(1 - \nu_s^2)]\ [1 - \nu_s\nu_c - (\nu_c - \nu_s)\cos^2\theta]$$

$$F_2 = [1/(1 - \nu_s^2)]\ [1 - \nu_s\nu_c - (\nu_c - \nu_s)/\cos^2\theta]$$

$$F_3 = (1 - \nu_c)/(1 - \nu_s)$$

If $\nu_c = \nu_s$, then $F_1 = F_2 = F_3 = 1$ and eq. (2.28) is simplified appreciably. In most cases, however, a mismatch in Poisson's ratio exists; and $\nu_c > \nu_s$, resulting in values of F_1, F_2 and F_3 that are computed from the equations above.

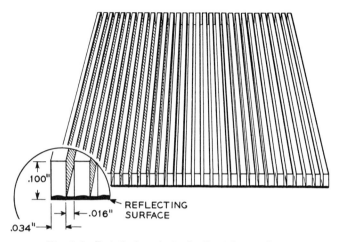

Fig. 2.9—Details for photoelastic strip coatings.

Strip Coatings[16]

A photoelastic strip coating is composed of closely spaced parallel strips as illustrated in Fig. 2.9. A coating where the strip thickness is several times the strip width does not react isotropically to strain because of its discontinuous geometry. The strips respond primarily to the strains in the specimen that are parallel to their long axes, and they are largely unresponsive to transverse and shear strains. To illustrate these strain-transmitting characteristics, consider the strain-optic law for a single strip oriented in the x direction:

$$N_s = S_{\|} \epsilon_{xx} + S_t \epsilon_{yy} + S_s \gamma_{xy} \qquad (2.29)$$

where N_s = fringe order exhibited by the strip
$S_{\|}$ = axial-strain sensitivity of the strip
S_t = transverse-strain sensitivity
S_s = shear-strain sensitivity

The values of the strain sensitivities depend upon the geometric proportions of the strip (principally the width w and height h_c for a strip that is long with respect to its cross-sectional dimensions). Strips with a large (w/h_c) ratio approach continuous coatings; and for this case $S_{\|} = -S_t = S_s = 2h_c/f_\epsilon$. However, when $w/h_c \ll 1, S_{\|} \longrightarrow 2h_c/f_\epsilon$; and S_t and S_s tend to vanish because the transverse and shear strains in the specimen are not transmitted into the coating.

Experiments by O'Regan[16] on strip coatings with $w/h_c = 0.34$ (see Fig. 2.9 for the dimensions of the strips) indicated that S_s was less than 2 percent of $S_{\|}$ and S_t was less than 1 percent of $S_{\|}$. Thus for strips with sufficiently small w/h_c, S_s and S_t can be neglected;

and a remarkably simple strain-optic law can be written as follows:

$$N_s = S_{\parallel} \epsilon_{\parallel} \qquad (2.30)$$

where $S_{\parallel} = 2h_c/f_\epsilon$. From eq. (2.30) it can be seen that the strips exhibit a photoelastic-fringe order proportional to the normal strain parallel to the strips. If the strips are positioned close together, the regions of extinction in adjacent strips blend and appear to form continuous fringes, as illustrated in Fig. 2.10. As a result the strip coatings provide whole-field data for the normal strain component parallel to the long axes of the strips.

Strip coatings can be employed in three different ways to establish the individual values of the principal stresses in the specimen. First, they can be used in the same manner as a strain-gage rosette, where three different strip coatings are used to measure the strains

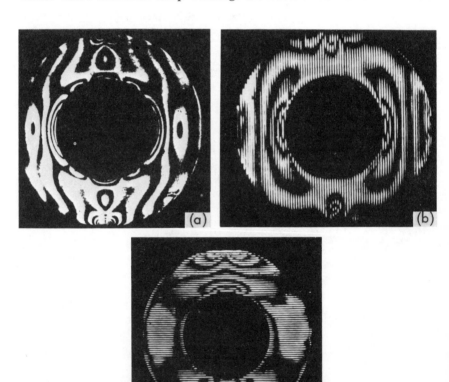

Fig. 2.10—Circular ring with (a) continuous coating; (b) strip coating, x direction; and (c) strip coating, y direction.

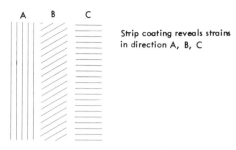

Fig. 2.11—Directions of the three strip coatings.

ϵ_A, ϵ_B, and ϵ_C, with the A, B, and C directions defined as shown in Fig. 2.11. The conventional rectangular-rosette equations can then be used to compute the principal stresses:

$$\sigma_1^s = E_s \frac{\epsilon_A + \epsilon_C}{2(1 - \nu_s)} + \frac{1}{2(1 + \nu_s)} \sqrt{(\epsilon_A - \epsilon_C)^2 + (2\epsilon_B - \epsilon_A - \epsilon_C)^2}$$

$$\sigma_2^s = E_s \frac{\epsilon_A + \epsilon_C}{2(1 - \nu_s)} - \frac{1}{2(1 + \nu_s)} \sqrt{(\epsilon_A - \epsilon_C)^2 + (2\epsilon_B - \epsilon_A - \epsilon_C)^2}$$

and the principal angle is given by

$$\tan 2\phi = (2\epsilon_B - \epsilon_A - \epsilon_C)/(\epsilon_A - \epsilon_B) \qquad (2.31)$$

The second method requires the use of two strip coatings to measure orthogonal strains (say ϵ_{xx} and ϵ_{yy}) and a continuous coating to measure $\epsilon_1 - \epsilon_2$. For this case

$$\epsilon_{xx} + \epsilon_{yy} = \epsilon_1 + \epsilon_2 = (f_\epsilon/2h_c)(N_{sx} + N_{sy}) \qquad (2.32)$$

where N_{sx}, N_{sy} are the fringe orders measured from the strip coatings in the x and y directions respectively. Assuming that f_ϵ and h_c are identical for all three coatings and solving eqs. (2.32) and (2.8) simultaneously for the principal strains,

$$\epsilon_1 = (f_\epsilon/4h_c)(N_{sx} + N_{sy} + N_0)$$

$$\epsilon_2 = (f_\epsilon/4h_c)(N_{sx} + N_{sy} - N_0) \qquad (2.33)$$

The principal stresses in the specimen are obtained from eqs. (2.33) and (2.26) as

$$\sigma_1^s = \frac{E f_\epsilon}{2h_c} \left[\frac{N_{sx} + N_{sy}}{2(1 - \nu_s)} + \frac{N_0}{2(1 + \nu_s)} \right]$$

$$\sigma_2^s = \frac{E f_\epsilon}{2h_c} \left[\frac{N_{sx} + N_{sy}}{2(1 - \nu_s)} - \frac{N_0}{2(1 + \nu_s)} \right]$$

and the principal angle is given by

$$\cos 2\phi = (N_{sx} - N_{sy})/N_0 \tag{2.34}$$

The third method of determining the individual principal stresses requires the use of only two coatings. In this procedure a strip coating is used to measure ϵ_{xx}, and a continuous coating is used to measure $\epsilon_1 - \epsilon_2$. The angle ϕ_1 between the principal-strain direction ϵ_1 and the x axis is determined from the isoclinic pattern of the continuous coating. These data can be substituted into the strain-transformation equations to give the principal strains:

$$\epsilon_1 = (f_\epsilon/2h_c)\,(N_{sx} + N_0 \sin^2 \phi_1)$$

$$\epsilon_2 = (f_\epsilon/2h_c)\,(N_{sx} - N_0 \cos^2 \phi_1) \tag{2.35}$$

The principal stresses are obtained from eqs. (2.35) and (2.26) as

$$\sigma_1^s = \frac{Ef_\epsilon}{2h_c}\left[\frac{N_{sx}}{1 - \nu_s} + \frac{N_0(\sin^2 \phi_1 - \nu_s \cos^2 \phi_1)}{1 - \nu_s^2}\right]$$

$$\sigma_2^s = \frac{Ef_\epsilon}{2h_c}\left[\frac{N_{sx}}{1 - \nu_s} - \frac{N_0(\cos^2 \phi - \nu_s \sin^2 \phi_1)}{1 - \nu_s^2}\right] \tag{2.36}$$

Although prospects for the future use of strip coatings appear bright, current usage is very limited since coatings are not commercially available in this form and application of such coatings to curved surfaces would present difficulties.

Numerical Integration

It is possible to employ the stress equations of equilibrium to separate the principal stresses in the coating if the coating is treated as a two-dimensional plane-stress problem in which the shear stresses acting at the interface surface of the coating vanish. If equilibrium of the principal element is considered, the Lamé-Maxwell equations of equilibrium can be expressed as

$$(\partial\sigma_1^c/\partial s) + [(\sigma_1^c - \sigma_2^c)/\rho_1] = 0 \qquad (\partial\sigma_2^c/\partial s^*) + [(\sigma_1^c - \sigma_2^c)/\rho_2] = 0 \tag{2.37}$$

where the curvilinear coordinates s_1 and s^* and the radii of curvature ρ_1 and ρ_2 are defined in Fig. 2.12.

The second term in each of eqs. (2.37) can be established from the isochromatic and isoclinic data. The individual value of σ_1^c, for example, can be obtained by a numerical procedure in which step-

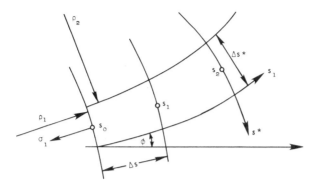

Fig. 2.12—Curvilinear coordinates for a principal element.

wise integration of eqs. (2.37) gives

$$\sigma_1^c]_{s_1} = \sigma_1^c]_{s_0} + (\sigma_1^c - \sigma_2^c)/\rho_1]_{(s_0+s_1)/2}\ \Delta s$$

$$\sigma_1^c]_{s_2} = \sigma_1^c]_{s_1} + (\sigma_1^c - \sigma_2^c)/\rho_1]_{(s_1+s_2)/2}\ \Delta s$$

$$\sigma_1^c]_{s_n} = \sigma_1^c]_{s_{n-1}} + (\sigma_1^c - \sigma_2^c)/\rho_1]_{(s_{n-1}+s_n)/2}\ \Delta s \qquad (2.38)$$

where $\qquad s_0$ = coordinate of a boundary point where σ_1^c is known

$s_1, s_2, \ldots, s_n$ = coordinates of closely spaced points along the s isostatic

$\qquad \Delta s = s_n - s_{n-1}$

This integration procedure is appreciably simplified in regions of the coating where either ρ_1 or ρ_2 goes to infinity. These regions are quickly identified in the isoclinic pattern as areas covered by a large, continuous-tone isoclinic of some arbitrary parameter. In such cases

$$\partial\sigma_1^c/\partial s = 0 \qquad \text{or} \qquad \partial\sigma_2^c/\partial s^* = 0$$

over the region considered. If $\rho_1 \longrightarrow \infty$, then σ_1 is a maximum or a minimum along s; and if $\rho_2 \longrightarrow \infty$, then σ_2 is a maximum or a minimum along s^*. Integration of eqs. (2.38) demonstrates that σ_1^c is either constant or changing linearly along s.

Electrical-resistance Strain Gages[17]

Electrical-resistance strain gages can be employed with photoelastic coatings to provide the additional information needed to separate the principal stresses. In this case the isochromatic and isoclinic data from the coating yield $\epsilon_1 - \epsilon_2$ and ϕ_1 over the entire strain field. The points of interest are selected, and a single-element strain gage is mounted at each such point with its primary sensing axis oriented in the ϵ_1 direction. The strain gages then indicate the principal strain, $\epsilon_1 = \epsilon_{sg}$; and from eq. (2.8)

$$\epsilon_2 = \epsilon_{sg} - Nf_\epsilon/2h_c \tag{2.39}$$

The principal stresses can then be obtained from:

$$\sigma_1 = E\left[\frac{\epsilon_{sg}}{1 - \nu_s} - \frac{\nu_s Nf_\epsilon}{2h_c(1 - \nu_s^2)}\right] \qquad \sigma_2 = E\left[\frac{\epsilon_{sg}}{1 - \nu_s} - \frac{Nf_\epsilon}{2h_c(1 - \nu_s^2)}\right]$$

$$\tag{2.40}$$

2.6 SUMMARY

The simplified approach to the theory of photoelastic coatings given here embodies the assumptions of perfect strain transmission from the specimen to the coating and uniform distribution of those strains through the thickness of the coating. With this idealized model of the coating behavior (applicable to many practical cases) the relationships between the principal stresses in the coating and the specimen are developed [eq. (2.5)]. Expressions for the mechanical behavior of the coating resulting from the transmitted stresses and strains are combined with the stress- and strain-optic laws to obtain the equations directly relating the optical response of the coating to the stresses and strains in the specimen [eqs. (2.7)–(2.13)].

The overall responsiveness of the coating is evaluated in terms of a stress-sensitivity index [eq. (2.14)] that combines both coating and specimen parameters influencing the optical output of the coating. The maximum available optical output of the coating (for elastic-stress analysis) is assessed and found equal to the product of the stress-sensitivity index and the yield strength of the specimen material [eq. (2.17)]. Maximum fringe orders achievable with a typical coating employed for elastic-stress analysis on a variety of engineering materials are presented in Table 2.1.

Techniques employed for fringe-order measurements—including whole-field methods with monochromatic and white light, and point compensation using the Babinet-Soleil compensator and Tardy compensation—are described in detail. Recommendations are also given for selecting the compensation technique on the basis of the resolution required and the range of fringe orders to be determined.

Several procedures are given for separating the stresses (i.e., obtaining the individual principal stresses). These include the oblique-incidence method [eq. (2.27)], strip coatings [eqs. (2.31), (2.34), and (2.36)], numerical integration [eq. (2.38)], and electrical-resistance strain gages [eq. (2.40)].

3
MATERIALS

Major developments in the application of two-and three-dimensional photoelasticity, as well as photoelastic coatings, occurred only after the introduction of suitable materials. Selection of the optimum photoelastic medium is essential in stress analyses.

3.1 PROPERTIES OF PHOTOELASTIC-COATING MATERIALS[18−21]

The demands on the mechanical and optical properties of photoelastic-coating materials are many; ideal polymers, perfectly suitable for coatings, do not exist. Although some excellent coating materials are commercially available, the physical properties of a nonexistent ideal material are listed below as a basis for comparison:

1. High optical sensitivity constant K to maximize the number of fringes per unit of strain for a given coating thickness.
2. Low modulus of elasticity to minimize reinforcing effects.
3. High resistance to both optical and mechanical stress relaxation for maximum stability of the fringe indication with time.
4. Linear strain-optic response for ease of data reduction.
5. Good bondability to typical engineering materials, ensuring perfect strain transmission at the interface with the coated object.
6. Strain-optic coefficient independent of temperature, minimizing problems in interpreting and correcting for temperature effects.
7. Negligible moisture absorption, eliminating consideration of parasitic birefringence caused by moisture-induced volume changes accompanying humidity changes.
8. Freedom from initial birefringence to avoid the necessity of correcting for this condition.

9. Good machinability for ease in fabricating complex configurations.
10. Comformability to curved surfaces, permitting application of the coating to the surfaces of three-dimensional components.

In most instances the selection of a coating material necessitates a compromise because no single material exists that exhibits all the foregoing properties. For example, the historically first coating material (employed by Mesnager[1]) was glass. While glass exhibits a relatively high sensitivity, it tends to cause significant reinforcement effects in the component because of its high elastic modulus. In addition, glass sheets cannot readily be conformed to curved surfaces, and are difficult to machine and cement to structures.

Numerous birefringent materials are available and can be employed as photoelastic coatings under appropriate circumstances. Typical properties of some of these materials are listed in Table 3.1. Examination of the list and comparison of the materials indicate that polycarbonate generally displays a superior combination of properties. Unfortunately, it is available only in sheet form, and techniques have not been developed for conforming or contouring the sheets to curved surfaces. Furthermore, the polycarbonate sheets are characterized by high internal stresses (with resultant initial birefringence), and the material is noticeably more difficult to bond than the epoxies, for example.

Table 3.1—Typical Properties of Various Materials Used as Coatings

| Material | Modulus E | | Sensitivity K | Strain Limit % |
	ksi	MPa		
Polycarbonate*	320	2,200	0.16	
PS-1†	360	2,500	0.15	10
Epoxy-anhydride‡	475	3,300	0.12	2
Epoxy-amine§	450	3,100	0.09	3
PS-2*†	450	3,100	0.13	3
Polyester‖	400	2,750	0.04	1.5
Modified Epoxy#	3	20	0.02	15
PS-3†	30	210	0.02	30
Polyurethane**	0.5	3	0.008	100
PS-4†	1.0	7	0.009	150
Glass	10,000	70,000	0.14	0.10

*Ref. 18.
†Commercially available from Photolastic, Inc., Malvern, Pa.
‡Ref. 16.
§100 pph ERL 2774, 15 pph TETA.
‖Homolite 100.
#Ref. 6.
**100 pph Hysol 2085, 24 pph Hysol 3562.

Epoxy photoelastic-coating materials also have comparatively high strain-optic coefficients. Since they are quite readily conformable and bondable, epoxy materials are generally preferred for applications requiring contouring. The procedures for applying epoxy coatings to curved surfaces are described in a later Section on "Curved Surfaces."

For measuring the large strains often encountered in plastic-strain studies, modified epoxies (a blend of rigid and flexible resins in a copolymer of epoxy and polysulfide) are usually employed as coatings since these materials can be compounded to exhibit a linear response over a large strain range (to 30 percent). In applications involving very large strains (in the range from 30 to 100 percent) polyurethane rubber can be used, since the low strain-optic coefficient of this material is no longer a serious handicap under these conditions.

The properties of the polymers listed in Table 3.1 can be varied widely by selection of the base resins and hardeners and by varying the relative proportions of the components. The material properties can also be altered by including additives such as diluents, modifiers, and plasticizers and by adjusting the curing process. The production of a polymeric sheet initially free of birefringence—with the proper sensitivity, elastic modulus, and strain capability and with optically acceptable surfaces—is not a trivial laboratory exercise. Some of the details involved in casting sheet materials are described in Section 3.4.

3.2 CALIBRATION METHODS

A photoelastic coating cannot be used for quantitative stress analysis unless its stress-optic or strain-optic sensitivity is known. To obtain this information requires calibration of the material, which is accomplished by subjecting a specified thickness of the material to a known stress state and measuring the photoelastic response with a polariscope. While a number of different stress states could be used for this purpose, uniaxial stress is probably the most common because it is easily achieved and the stress and strain distributions are accurately predictable from simple analytical relationships.

Beam in Bending

Calibration of high-sensitivity photoelastic-coating materials such as the polycarbonates and the rigid epoxies is usually accomplished by employing a cantilever beam in bending. The coating is placed on one side of the beam, and the beam is strained by dead weights F or a predetermined deflection y_0 applied at the free end. The difference between the principal strains at some point A at a distance L from the free end of the beam is

$$\epsilon_1^s - \epsilon_2^s = 6 \, (1 + \nu_s) \, LF/b_s h_s^2 E_s \qquad (3.1)$$

or

$$\epsilon_1^s - \epsilon_2^s = 3 \, (1 + \nu_s) \, h_s L y_0 / 2 l_s^3 \qquad (3.2)$$

where b_s, h_s, and l_s are the width, height, and effective length of the beam respectively (see Fig. 3.1).

Since the strains through the thickness of the coating vary linearly, a correction factor C_2 [see eq. (5.9)] is used to establish the average strain difference through the thickness as

$$(\epsilon_1^c - \epsilon_2^c)_{\text{ave}} = C_2 \, (\epsilon_1^s - \epsilon_2^s) \qquad (3.3)$$

Substituting eqs. (3.1), (3.2), and (3.3) into the strain-optic law [eq. (2.8)] gives the material-fringe value in terms of strain as

$$f_\epsilon = \lambda/K = 12(1 + \nu_s) \, h_c C_2 LF/N b h_s^2 E_s \qquad (3.4)$$

for the dead-weight loading F and

$$f_\epsilon = \lambda/K = 3(1 + \nu_s) \, h_c C_2 h_s L y_0 / N l_s^3 \qquad (3.5)$$

for the application of the free-end deflection y_0 (assuming that the length of the coating strip is small and that the coating is not influencing the deflection y_0).

As an example, consider a calibration beam 0.25×1.0 in. (approx. 6×25 mm) in cross section, with an effective length of 12 in. (305 mm). The beam is fabricated from 2024-T4 aluminum and has a coating that is 0.105 in. (2.67 mm) thick applied to one side. Taking the distance L to the point of the fringe-order reading N as 8 in. (203 mm) and the weight F as 10 lb (44.5 N), from eq. (3.1) the difference in principal strains is 982 μin./in. (μm/m), (assuming $\nu_s = 0.33$ and $E = 10.4 \times 10^6$ psi (70,000 MPa). The correction factor C_2 is established as 1.25 from Fig. 5.4 for a thickness ratio $hc/hs = 0.42$.

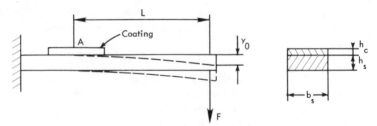

Fig. 3.1—Calibration beam.

Accurate measurement of the fringe order N with a reflection polariscope, employing an appropriate compensation method, is essential. Usually, several readings of N are taken as the load is increased in, for example, 20 percent increments to the maximum load. A graph of F versus N is plotted, and a best-fit straight line is constructed through the data points. The slope of this line (in this example, 0.163 fringes/lb or 0.0366 fringes/N) gives the best value of N/F to be used in eq. (3.4) for determining the calibration constants f_ϵ and K. For this example, $f_\epsilon = 158 \times 10^{-6}$ in./fringe (4.01×10^{-6} m/fringe) and $K = 0.143$, assuming $\lambda = 22.7$ μin. (577 nm) for white light.

A similar procedure is used when the calibration is performed with a predetermined free-end deflection of the beam. In this instance, the value of the slope y_0/N is accurately established and employed in eq. (3.5) to obtain f_ϵ and K.

Tension Specimen

Calibration of high-sensitivity coatings with a beam in bending is usually preferred because of the simplicity of this method. However, the strains developed in the beam are generally too small to produce an adequate optical response in low-sensitivity coatings. Because higher strains (10 to 100 percent) are required to calibrate the relatively insensitive coating materials, the calibration is best performed on the material itself. A tension specimen of the polymer (unbonded) is loaded in uniaxial tension, and both the axial strain and the fringe order are measured. The axial strain ϵ_1^c is easily determined with an extensometer since the strain is large. The fringe order can be measured with either a reflection or transmission polariscope. From eq. (2.8) it can be seen that the coating calibration constants can be established in terms of the data obtained from the tension specimen as

Transmission polariscope: $f_\epsilon = \lambda/K = (1 + \nu_c)\, h\, (\epsilon_1^c/N)$

Reflection polariscope: $f_\epsilon = \lambda/K = 2(1 + \nu_c)\, h\, (\epsilon_1^c/N)$ \qquad (3.6)

Other specimen configurations and stress states can also be used for calibration if the principal-strain differences at a point can be predicted accurately. The circular disk, quite commonly used as a calibration specimen in transmission photoelasticity to determine f_σ, is less effective in the establishment of f_ϵ. Although the material fringe values f_σ and f_ϵ are related through eq. (2.10), the accurate conversion from the stress-fringe to the strain-fringe value requires precise knowledge of the mechanical properties of the coating material, i.e., E_c and ν_c. Practical difficulties in accurately determining E_c and ν_c restrict the use of the disk for measuring f_ϵ.

3.3 SELECTION OF COATING THICKNESS

After the type of coating is selected (i.e., polycarbonate, epoxy, polyurethane, etc.), the thickness must be specified. Two major (and conflicting) considerations—coating sensitivity and thickness effects—must be weighed in specifying the coating thickness. The coefficient of sensitivity (defined in Section 2.3) varies linearly with the thickness, i.e., $C_c = 2h_c/f_\epsilon$. Thus the optical output of the coating is proportional to the thickness. However, as h_c is increased, thickness effects such as reinforcement and strain variation through the thickness of the coating may become significant and require consideration (see Chapter 5).

Another factor of equal importance is the accuracy with which the fringe order N can be measured. The absolute error in the difference between the principal strains can be written as

$$\Delta(\epsilon_1^c - \epsilon_2^c) = (f_\epsilon/2h_c)\Delta N = \Delta N/C_c \qquad (3.7)$$

where ΔN is the error in measuring N. From this it is apparent that increasing the coating thickness decreases the absolute error $\Delta(\epsilon_1^c - \epsilon_2^c)$. However, this error can also be decreased by minimizing ΔN. As discussed in Section 2.4, full-field observations of colored fringe patterns give $\Delta N = \pm 0.1$ fringe. Point-by-point compensation methods, on the other hand, can be used to reduce ΔN to ± 0.01 fringe.

The relative error e can be expressed by

$$e = \Delta(\epsilon_1^c - \epsilon_2^c)/(\epsilon_1^c - \epsilon_2^c)_{\max} \qquad (3.8)$$

where $(\epsilon_1^c - \epsilon_2^c)_{\max}$ is the maximum principal-strain difference occurring over the field.

From eqs. (3.7) and (3.8) the relative error can be written as

$$e = f_\epsilon \Delta N/2h_c(\epsilon_1^c - \epsilon_2^c)_{\max} = C_c \Delta N/(\epsilon_1^c - \epsilon_2^c)_{\max} \qquad (3.9)$$

This relationship shows that the relative error is also affected by $\epsilon_1^c - \epsilon_2^c$, which should be maximized to minimize e. Equation (3.9) can now be rewritten in a form permitting an estimation of the necessary coating thickness to satisfy a prescribed accuracy requirement e:

$$h_c = f_\epsilon \Delta N/2e(\epsilon_1^c - \epsilon_2^c)_{\max} \qquad (3.10)$$

As an example, assume that a stress analysis is to be performed in the elastic range on a component fabricated from 2024-T4 aluminum alloy.   The value of $(\epsilon_1^c - \epsilon_2^c)_{max}$ must be below that corresponding to yielding, or less than 0.005.   Assume that the selected coating has the properties $K = 0.09$ and $f_\epsilon = 250 \times 10^{-6}$ in./fringe (6.25 $\times$ 10^{-6} m/fringe).   Suppose that full-field readings of colored fringes observed in white light give $\Delta N = \pm 0.1$ and that the relative error e is to be less than 5 percent.   Direct substitution into eq. (3.10) indicates that the coating should be at least 0.050 in. (1.27 mm) thick.   Alternatively, fringe-order determinations could be made using compensation techniques, with $\Delta N = \pm 0.01$ and with the requirement that the relative error be less than 1 percent.   In this case, evaluation of eq. (3.10) shows that the coating should be at least 0.025 in. (0.63 mm) thick.

Following the original estimate of coating thickness as obtained from the foregoing procedure, the thickness must be adjusted to accommodate other considerations.   For thin plates, data analysis is simplified by selecting h_c/h_s for equal magnitudes of the thickness-effect correction factors (i.e., $C_1 = C_2$ —see Figs. 5.2 and 5.4).   When this is done, C_1, which accounts for the plane-stress reinforcing effect, cancels C_2, which corrects for thickness effects due to bending.

The selection of very thin coatings (e.g., less than 0.020 in., or 0.05 mm) is not recommended for precise stress analysis since the relative error introduced by the variations in thickness from point to point can become excessive.   Very thin coatings are usually employed in qualitative observations such as the detection of yielding, proof testing, or locating the areas where the stress is maximum and determining the principal-stress directions.

3.4 APPLICATION OF COATINGS

Once the type of coating has been selected and the thickness specified, the next step in a stress analysis using photoelastic coatings consists of bonding the coating in place on the test structure or component.   Coating applications can be broadly divided into two classes according to shape.   The surface of the specimen may be either plane or curved.   Because the application procedure is markedly different for these two cases, they are treated separately in the following sections.

Plane Surfaces[22]

Application of photoelastic coatings to plane surfaces can be achieved most readily and economically by cementing flat, precured sheets of material to the surface.   The successful use of photoelastic coatings depends upon a perfect bond between the coating and the specimen.   For this reason, careful attention must be given to all aspects of the application process, including proper surface preparation and cleaning, preparation of the polymeric coating and cement, and bonding procedure.

Preparation of the specimen surface consists first of the removal of all foreign matter, followed by sanding, and finally cleaning with suitable chemical agents and solvents.   The flat sheet of coating is machined to size, cleaned, properly located and aligned relative to the specimen, and then bonded.   As an alternative, the sheet can be bonded to the specimen, which then can be used as a template to machine the coating to the proper shape.   When feasible, the first procedure is preferred because it avoids the risk of bond damage due to the mechanical and thermal stresses generated in the machining process.

As illustrated in Fig. 3.2, two different types of coating boundaries can exist in plane-surface coating applications. A discontinuity exists along edge A-A' in that the specimen surface extends beyond the coating.   At this edge, large forces are applied to the coating by shear stresses acting at the interface.   As a result, meaningful fringe measurements cannot be made in the region close to this edge (for a distance of about four times the thickness of the coating). The finish of such an edge is not critical, and the normal practice is to bevel the edge or to provide a fillet formed with cement to eliminate the high stress concentration at the discontinuity between the coating and the specimen.   Both of these edge treatments are illustrated in Fig. 3.2.

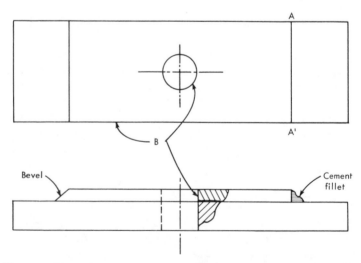

Fig. 3.2—Type of boundaries between the coating and test specimen.

Location B represents a discontinuity in the test specimen; the coating boundary must be precisely matched to the specimen boundary. The measurements in this region are generally most significant, and the edges of the coating should be machined carefully so that they are perpendicular to the surface and free of residual stress or birefringence. If any initial birefringence is present in the coating, it can be reduced or eliminated by annealing. However, the annealing must precede the cementing operation. During cementing, care must be exercised to align the discontinuity in the coating (here, a hole) with the one in the specimen so that the boundaries match.

The cement, usually an aluminum-filled epoxy, is spread evenly over the test area, and the sheet of coating is applied. Commonly, the coating sheet is placed down at one edge and slowly rotated and pressed into position, working out the air bubbles in the process. Pressure is not ordinarily applied to the sheet while the cement is curing. The curing is accomplished at room temperature over a 24-hour period. When the adhesive is cured, the thickness is in the range of 0.003 to 0.010 in. (0.08 to 0.25 mm).

Curved Surfaces[22]

Photoelastic coatings may be applied to the curved surfaces that are found in many test components by using a contoured-sheet technique. This technique is illustrated in Fig. 3.3 where the four primary operations in the application sequence are described. First, a sheet of flexible epoxy is cast in an open-face mold. The mold is a Teflon-covered surface plate that is carefully leveled to insure a uniform thickness of the cast sheet. The epoxy is formulated with a relatively slow-curing agent so that the cast sheet slowly transforms from a liquid (A stage) to a rubbery (B stage) sheet.

While the sheet is still soft and pliable, it is stripped from the open-face mold and then contoured to fit the curved surface of the test specimen. The deformations imposed on the sheet in the contouring process do not produce significant birefringence in the coating since the optical sensitivity constant is very low when the epoxy is in the B stage. The sheet, now contoured, is maintained in contact with the curved surface of the test specimen until polymerization is complete and the epoxy is cured (C stage). The contour sheet is rigid and may be removed to be cleaned, measured for any thickness changes, and trimmed to final size. The contoured shell is then cemented back in place on the curved surface of the test specimen.

The contoured-sheet technique permits the application of sensitive photoelastic coatings to components of almost any shape. The coating is formed into shells of nearly uniform thickness, which con-

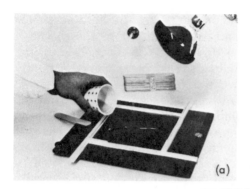

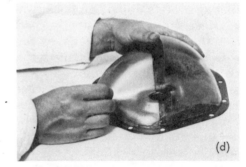

Fig. 3.3—Sequence of operations in the formation of contoured sheets: (a) Pour
liquid resin on a flat plate; (b) before full hardening, lift the plastic
from the plate; (c) shape the plastic to fit the contour of the test part
and allow to harden; and (d) cement the plastic shell to the test part.

form closely to the surface of the component. In practice, experience and skill is necessary in applying contoured coatings, and care must be exercised in each step of the operation.

The design of the resin system for contoured-sheet applications is much more difficult than formulating resins for casting flat sheets. The materials must (1) be very pliable in the B stage of the cure, (2) develop photoelastic response only when completely polymerized, (3) exhibit low viscosity for ease in mixing at room temperature, and (4) polymerize within a reasonable time at room temperature.

Cements

Cements for photoelastic coatings must be designed for compatibility with the particular coating employed; and since an excellent adhesive bond is essential to the method, the cement should be carefully selected, mixed, applied, and cured. An adequate cement should exhibit the following characteristics:

1. Long pot life, to allow sufficient time for application.
2. Low viscosity, permitting excess adhesive to flow from under the applied sheet.
3. Short total curing time at room temperature.
4. Low stress-relaxation tendency after cure.
5. Modulus of elasticity greater than or at least equal to that of the coating.
6. Sufficient strain capability to transmit strains up to the maximum elongation of the coating.
7. Uniform and high reflectivity.

For the high-sensitivity/low-elongation coatings, amine-cured epoxies can be formulated to meet the foregoing requirements. Generally, an addition of 10 to 30 percent, by weight, of aluminum powder is made to the cement so that the bonding layer becomes reflective. The aluminum powder also enhances the stress-relaxation properties of the cement and improves its handling characteristics.

When clear adhesives are used, the bonding layer contributes to the optical response of the coating. In the case of thin coatings bonded with clear adhesives, the strain-optic coefficient and the thickness of the bond layer become important; and their effects must be taken into account. Furthermore, the surface of the test object must be polished to give it sufficient reflectivity for use with a reflection polariscope. For these reasons, cements filled with aluminum particles are usually preferred.

Coatings such as the modified epoxies and the polyurethanes, which exhibit large elongations, must be bonded with cements capable of equally large strains. It is often advisable to use the liquid counterparts of the sheet materials as the adhesives for these large-elongation coatings.

3.5 APPLICATION OF LIQUID COATINGS

Coatings in the form of liquid resins can be applied directly to the surface of a test member by dipping, spreading, brushing, or spraying. Thin coatings can be effectively applied to plane surfaces by preheating the surface to 120–140°F (50–60°C) and brushing on a coating composed of epoxy resin and amine hardener. Polymerization can be achieved in a relatively short time, and the member can be tested within a few hours. Thicker coatings can be built up by applying successive layers, but cementing precured flat sheets onto the test surface is the preferred method of obtaining thicker coatings. One of the problems in using a thin coating applied as a liquid is the variation in the coating thickness from point to point. These variations are likely to be relatively large compared to the average thickness of the coating. The thickness must be determined at each point where a fringe order is measured, and this requires magnetic, capacitive, optical, or other gages.

Applications of liquid coatings to large curved surfaces can be accomplished most readily by spraying. Thin coatings applied in this manner can be used to detect yield regions after the proof loads are removed from a structure. The most practical approach involves mixing the epoxy resin with an amine hardener and a solvent. The solvent improves the spraying characteristics of the mixture and tends to increase pot life. After spraying, the solvent evaporates, and the resin polymerizes on the test surface. This can be an economical method since large surfaces can be coated in a relatively short time.

3.6 SUMMARY

Major developments in the application of photoelastic coatings occurred after the introduction of high-sensitivity, low-modulus photoelastic materials. Of the many polymeric materials, polycarbonate is generally considered superior; however, it cannot be applied to surfaces with compound curvatures. Epoxies are preferred for applications requiring contouring of the coating. Finally, copolymers of epoxy and polysulfide or polyurethane are used in applications where large strains are encountered.

Coatings are calibrated on cantilever beams with an applied load

F to obtain f_ϵ as

$$f_\epsilon = 12(1 + \nu_s)h_c\,C_2\,LF/Nbh_s^2\,E_s$$

or with an applied free-end deflection y_0 where

$$f_\epsilon = 3(1 + \nu_s)h_c\,C_2\,h_s\,Ly_0\,/Nl_s^3$$

When a coating is calibrated in tension,

$$f_\epsilon = 2(1 + \nu_c)hP/AEN$$

The coating thickness required to limit the error e is

$$h_c = f_\epsilon\,\Delta N/2e\,(\epsilon_1^c - \epsilon_2^c)_{\max}$$

when an error of ΔN in fringe-order measurement is anticipated.

4 INSTRUMENTS

Three independent parameters must be measured or otherwise known to completely determine the state of stress or strain at a point on the surface of a test member. This requirement applies to all methods of experimental stress analysis, including the photoelastic-coating method. Because the photoelastic phenomenon is primarily responsive to the principal stresses or strains and their directions, the photoelastic-coating method is ordinarily used to measure ϵ_1, ϵ_2, and ϕ_1, the isoclinic angle defining the principal stress direction (or σ_1, σ_2, and ϕ_1). In some cases one or two of these parameters may be known from other considerations, reducing correspondingly the number of independent measurements that must be made. Consider, for example, a plane-stress state including a free boundary. Everywhere on such a boundary $\sigma_2 = 0$, and the principal axes are tangent and normal to the free boundary. Under these circumstances only a single measurement is required to completely define the states of stress and strain.

The measuring instrument used with a photoelastic coating is the reflection polariscope. With normal light incidence, the instrument indicates the directions of the principal axes by the isoclinic-fringe pattern. Isochromatic fringes that are measured under normal-incidence light provide the difference between the principal strains. Under oblique incidence the isochromatics can be used to measure a third parameter, i.e., the difference between the secondary principal strains. Thus the reflection polariscope (with the proper auxiliary equipment) can be used to measure all three independent parameters required to completely define a generalized biaxial state of stress or strain (i.e., the state of stress or strain on the free surface of any arbitrarily shaped body under any arbitrary mode of loading).

Reflection polariscopes for use with photoelastic coatings are based upon the same optical principles as the transmission polariscopes used for two-dimensional models. The primary difference is that in the reflection polariscope the light path is folded at the reflective interface between the coating and test member and thus traverses the coating thickness twice. In addition, reflection polariscopes are usually designed for portability so that they can be used to make field measurements on structures that cannot conveniently be brought into the laboratory. Many specialized accessories have been developed for reflection polariscopes, including those for (1) dynamic stress analysis (stroboscopic lights, motion-picture cameras, and photosensitive detection and recording equipment); (2) remote measurements (telescopes and television cameras); (3) magnified close-up measurements (microscopes); (4) precise point-by-point measurements (compensators); and (5) measurements of secondary principal-strain differences (oblique-incidence attachments). With the appropriate auxiliary equipment, the reflection polariscope can obtain quantitative measurements over an extremely wide range of operational and environmental conditions.

4.1 REFLECTION POLARISCOPES FOR NORMAL INCIDENCE

Several arrangements of reflection polariscopes are shown in Fig. 4.1. The system illustrated in (a), which uses the partial mirror, was proposed early in the development of the photoelastic-coating method. It permits perfectly normal incidence of light but its use is handicapped because the light reflected from the first surface of the coating is congruent with the light that traverses the coating, thus degrading the fringe contrast. Furthermore, passage of the light through the partial mirror alters the ellipticity of polarization. While this effect is inconsequential in the plane polariscope (observing isoclinic fringes), it is deleterious when circular polarization and compensation are used. There is also a tendency for instruments of this type (although limited to a small field of observation) to be too large, heavy, and cumbersome for portable use.

Because of the disadvantages already noted, the system shown in Fig. 4.1(a) is often abandoned in favor of the more practical approaches illustrated in Figs. 4.1(b), (c), and (d). The arrangement in Fig. 4.1(c), with the quarter-wave plates laminated to the polarizer/analyzer, has only very limited application since isoclinics cannot be observed directly nor can compensation be employed. Sketch (d) shows a concentric polarizer and analyzer, with annular illumination. This system has been described by Slot[23] and is capable of excellent

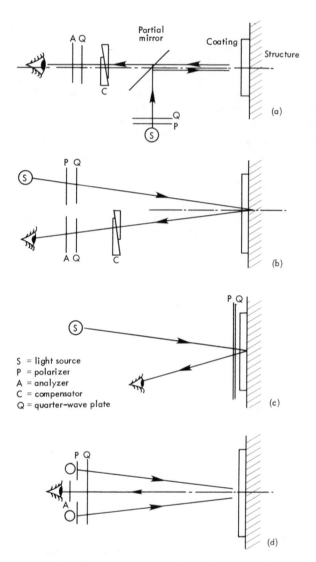

Fig. 4.1—Possible arrangements for a reflection polariscope.

results, although it is not commercially available. The most popular reflection polariscope is shown in Fig. 4.1(b). In this arrangement the polarizer and analyzer are separated, and the light incidence is not perfectly normal. The error introduced by this condition is negligible, however. Assuming typically a center-to-center distance between the polarizer and analyzer of 5 in. (127 mm), with the instrument 2 ft (0.6 m) distant from the coated test specimen, the

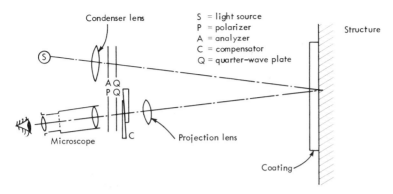

Fig. 4.2—Optical system in a polariscope.

oblique angle θ in the coating is less than 4 deg, causing an error of less than 1 percent.

As shown in Fig. 4.2, this type of polariscope is usually equipped with an optical system to provide adequate magnification (5× to 20×) for observation of details in the fringe pattern. Since photoelastic coatings are applied to actual structures, the stress measurements are often performed under difficult field conditions; and the overall polariscope design must be sufficiently versatile to function effectively in such circumstances. Polariscopes of this type are generally intended for either tripod mounting or hand-held use. Fig. 4.3 illustrates a modern reflection polariscope mounted on a tripod and equipped with a magnifying arrangement and a camera.

4.2 OBLIQUE-INCIDENCE POLARISCOPE

One method of achieving an oblique-incidence polariscope is illustrated in Fig. 4.4(a), which represents a special-purpose unit based upon an equilateral glass prism and employs a 45-deg angle of incidence. In this case the index of refraction of the glass is matched to that of the photoelastic coating. A drop of oil between the glass and the coating provides the necessary optical continuity. This design is limited in application to fringe-order measurements on flat or slightly concave surfaces. The system is rather inconvenient in practice and can introduce errors when used in testing lightweight structures because the weight of the instrument may add to the applied load when it is placed in contact with the structure.

The system shown in Fig. 4.4(b) is usually preferred because it is simply an attachment to a conventional normal-incidence reflection polariscope and because physical contact with the coating is eliminated. Avoidance of physical contact with the coating limits the

Fig. 4.3—Reflection polariscope equipped with a magnifying arrangement and
       camera.

angle of incidence that can be used. The maximum angle of incidence obtainable in this arrangement is defined by:

$$\sin \theta_{max} = 1/n_0 \qquad (4.1)$$

which yields θ_{max} = 39 deg for n_0 = 1.6. In practice, however, as the
angle of incidence increases, the fraction of the incident light reflected off the coating surface also increases, and the fringe contrast
is degraded.

In deriving the relationships for oblique incidence given in the
section on "Oblique Incidence" in Chapter 2, it was assumed that the
direction of observation was rotated about the principal stresses σ_1.
This condition requires that the directions of the principal stresses
should be established before oblique-incidence isochromatic measurements are attempted.   In practice, the directions of the principal
stresses are determined by removing the quarter-wave plates from the
polariscope and crossing the axes to obtain a dark field. The entire
instrument is then rotated while observing the isoclinic pattern.
When an isoclinic fringe coincides with the point under consideration, alignment with σ_1 and σ_2 is achieved. Insertion of a compensator in the field determines the direction σ_1. The direction of ob-

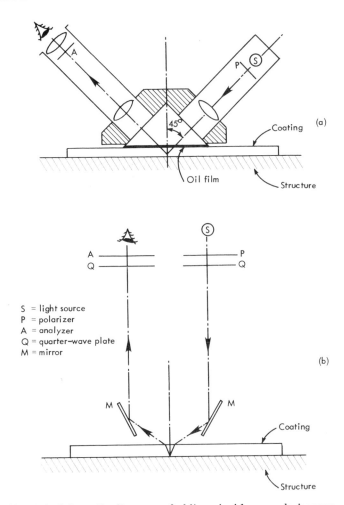

Fig. 4.4—Schematic diagrams of oblique-incidence polariscopes.

servation can then be established to give an oblique incidence angle θ by rotation about the σ_1 direction.

4.3 MEASUREMENTS, INCLUDING COMPENSATION METHODS

In the following sections practical techniques are described for making stress measurements on photoelastic coatings with a reflection polariscope and the appropriate auxiliary equipment. The discussion covers the two basic photoelastic measurements; i.e., the directions of the principal stresses and the fringe orders. Particular attention is paid to techniques for measuring fractional fringe orders with compensators since these accessories are widely used with reflection polariscopes. Photoelastic coatings on common structural materials often produce low fringe orders (see Table 2.1). Compensation methods are therefore necessary to achieve sufficient resolution in fringe-order measurement.

Directions of Principal Stresses

The directions of the principal stresses are obtained directly from observation of the isoclinic fringes. The isoclinic parameters are explored and recorded in the same general manner as in a conventional transmission polariscope, i.e., by one of the three following methods:

1. Photographs of the isoclinic fringes are taken at different angles from the reference axis in, for example, 10-deg increments.
2. The isoclinics are traced directly on the coating and then photographed after identifying the angular parameter.
3. Point-by-point measurements are made and the results used directly in fringe-order measurements at the same points.

Note that the measurement of directions are independent of the coating thickness h_c.

Modern reflection polariscopes ordinarily are designed so that the polarizer and analyzer can be rotated synchronously while maintaining their optic axes crossed for isoclinic exploration. When the unit is used as a plane polariscope for isoclinic observation, the quarter-wave plates are either physically removed from the light path or oriented in such a fashion that they do not alter the linearly polarized light. In the latter case the optic axes of the quarter-wave plates are simply aligned with the crossed polarizer and analyzer axes. Isostatic lines are constructed from the complete isoclinic diagram with the same procedure as used in classical two-dimensional photoelasticity.[10,12,24]

Direct Isochromatic Measurements

As in two-dimensional model photoelasticity, integral and half-order fringes can be counted in either white or monochromatic light. For integral fringe orders in white light, the tint-of-passage fringes are counted. These are the sharp dividing lines located between red (or pink) and blue (or green). There is a preference for using the tint of passage because (1) it provides maximum sensitivity in the observation of its positioning, and (2) the tint-of-passage fringes are proportional to $\epsilon_1 - \epsilon_2$, as indicated in Table 2.2.

With photoelastic coatings, fringe counting is ordinarily used as a measurement technique only for large strains. For many materials, strains in the elastic regime are small (see Table 2.1), and only a few fringes can usually be observed. Consequently, fractional fringe orders must be measured. A simple technique for this is by color comparison, which consists of comparing isochromatic fringe colors with a color chart developed by calibration (see Table 2.2). A further improvement in accuracy can be achieved when isochromatic color fringes are matched with a calibrated frozen-fringe specimen observed through the same polariscope as the part under investigation. For this purpose the frozen-fringe standard can take a variety of forms such as a beam in bending or a tapered tensile specimen. The resolutions obtainable with these methods are approximately ± 0.25 fringe for color comparison and ± 0.15 fringe for color matching. This level of resolution is often inadequate for elastic stress analyses, and compensation techniques are necessary. For the study of plastic deformations or for general assessment of the order of magnitude of the stresses present, color matching is very convenient.

Compensators

General. Compensators are instruments for measuring fractional fringe orders. There are two basic approaches to compensation. In the first, compensation is achieved by converting the elliptically polarized light emerging from the photoelastic medium into linearly polarized light and then extinguishing this light by rotation of the analyzer. The angle of rotation of the analyzer from its initial dark-field position is proportional to the fractional fringe order being measured. This method of compensation is known as the Tardy method, and similar techniques have been developed by Senarmont and others.

With the second approach to compensation, another birefringent element is inserted in series with the photoelastic model. The birefringence in this element can be adjusted to provide a retardation equal in magnitude and opposite in sign to that occurring in the photoelastic model. When the adjustment of the compensating element is complete, the net birefringence in the light path is zero and the point under consideration is observed as black in the field of the polariscope.

Several different compensators have been developed and are described below:

1. The Berek compensator is simply a birefringent plate with a uniform thickness. The adjustment of the birefringence is made by

rotating the plate so that it is oblique relative to the direction of propagation of the light. This rotation increases the distance the light must travel in passing obliquely through the plate and increases the birefringence. Also, a change in birefringence occurs because of the different indices of refraction on the plane of the plate, which is perpendicular to the oblique light. Since the birefringence exhibited by the plate is not linear with the angle of rotation, a calibration curve is required for interpretation.

2. The Babinet compensator employs a quartz wedge, the birefringence of which varies linearly with the length of the light path from the point of entry. Thus the birefringence is proportional to the wedge thickness traversed by the light. The Babinet compensator is a linear device, but it has the disadvantage that a continuous birefringent gradient (with many fringes) is always visible in the field. While this characteristic makes reading the instrument very convenient and accurate in areas of uniform strain (constant birefringence in the photoelastic medium), it makes readings more difficult in regions where there are strain gradients (the common practical case).

3. The Babinet-Soleil compensator is similar to the Babinet compensator but uses two quartz wedges, as shown in Fig. 2.5. Within the field of observation the birefringence is uniform. Change in birefringence is accomplished by displacing one wedge with respect to the other, producing a corresponding change in the total thickness of the assembly and in the length of the light path. However, the thickness of the assembly is always uniform, and thus uniform birefringence is observed in the field of view. The Babinet-Soleil compensator is equally effective in regions of low- or high-strain gradients and is very convenient to use because of its intrinsic linearity.

4. A variation of the preceding method, compensation by temporary birefringence, can be achieved by inserting into the optical path a photoelastic specimen that is subjected to a known axial load (e.g.) tension or compression and by varying the load until the observed total birefringence is zero. If the plastic in the specimen is of the same type and thickness as the photoelastic medium under investigation, the stress in the compensating medium is the same in magnitude and opposite in sign to that in the test specimen when zero net birefringence is observed.

Compensators for Photoelastic Coatings. Although all the compensation methods that have been described can be employed with photoelastic coatings, two of these techniques are commonly used in actual practice.

Tardy compensation can be used whenever fractional fringe orders are required and is particularly useful for full-field measurements. The Tardy method is fast and convenient but has the disadvantage that it identifies only the fractional fringe order, and supplementary means must be used to establish the integral order to which the fraction should be added (algebraically). The integral fringe order can be found by counting fringes from a known zero-order fringe or by counting and identifying fringes as the part is loaded from a stress-free condition. Fractional fringe orders can be resolved to within ±0.02 to ±0.05 fringe with Tardy compensation, depending upon the experience of the operator. A complete description of the Tardy method is presented in Section 2.4.

A commercial form of the Babinet-Soleil compensator uses two juxtaposed frozen-stress photoelastic strips instead of two quartz wedges to cancel or compensate for the birefringence in the coating (Fig. 2.5). The locked-in birefringence is linearly distributed along the length of the strips, and the gradients are in opposite directions. Thus the total birefringence introduced by two strips in series is uniform along its length and proportional to the displacement of one strip with respect to the other. The sign of the strip displacement (positive for increasing birefringence) is used to determine the direction of the algebraically maximum principal stress σ_1 in biaxial stress fields, or the sign of the nonzero principal stress in a uniaxial stress field. Suppose, for instance, that the compensator is aligned with one of the principal stresses and it is found that an increase in birefringence is necessary to cancel the optical response. This implies that the compensator is aligned with the σ_2 direction.

In another commercial form of the Babinet-Soleil compensator, both the magnitude and sign of the displacement are directly indicated by a mechanical counter that is coupled to the screw and translates the strip. In an alternative arrangement a linear potentiometer is used to convert the strip displacement into an electrical signal proportional to the birefringence. The electrical output is digitally displayed in the form of the difference in principal strains or stresses in any desired engineering units. The same information can also be printed to obtain a permanent record. A contemporary instrument incorporating many of these features is shown in Fig. 4.5 and is capable of measuring fringe orders to ±0.01 fringe.

4.4 CALIBRATION

Two calibrations are required in stress measurement with photoelastic coatings. One of these is for the coating thickness h_c, and the

Fig. 4.5—Reflection polariscope equipped with a digital compensator and
auxiliary instrumentation.

other is for the strain-sensitivity constant K. The thickness measurement can be made with a micrometer on either flat or contoured
sheets prior to cementing the sheets to the structure. The thickness
will usually be constant to within a few thousandths of an inch over
the entire surface of the sheet. Significant deviations, if any, should
be recorded, and their locations marked. Since reflective cements are
normally used for bonding, the thickness of the cement does not
directly alter the results. A correction for cement thickness may be
required, however, when the coating is bonded in place with a transparent epoxy cement. When coatings are applied by brushing or
spraying, the coating thickness must be measured directly on the
part. This can be done with a calibrated microscope by focusing first
on the surface of the test part proper and then on the outer surface
of the coating. The displacement of the microscope necessary to
focus separately on these two surfaces is a measure of the coating
thickness.

Calibration for the strain-sensitivity constant K can be accomplished with any device capable of subjecting a coated specimen to a
known strain. The strain is substituted into Eq. (2.9) to solve for K.
Calibration specimens subjected to in-plane loads usually require the
availability of a universal testing machine. The loads applied to the

calibration specimens fabricated from metallic alloys are large when strains approaching the yield strain of the material are applied.

To reduce the loads necessary for calibration and to eliminate the requirement of a testing machine, calibration specimens involving either plates or beams in bending are recommended. Simple manually actuated cantilever beam bending fixtures can be constructed that generate high strains in beams on the order of 10 in. (254 mm) in length. Calibration procedures utilizing a cantilever beam are presented in Section 3.2.

The primary inconvenience associated with this method of calibration is that bending introduces two (determinate) errors in the coating indication, which must be corrected for. The first of these is due to the reinforcing effect of the coating on the beam; and the second arises from the fact that the optical indication of the coating corresponds to the average strain through the thickness of the coating, rather than the strain at the surface of the metal beam. Since these errors are determinate and predictable, the necessary correction factors can be calculated for any specified set of circumstances. The appropriate correction factors are normally supplied by the manufacturer of the calibration fixture in convenient graphical form. The calibration fixture can be of either the controlled-load or controlled-deflection type, as long as calibration specimens do not exhibit creep or stress relaxation with time. The fixture correction factor is different, however, for each case. Fig. 4.6 illustrates a controlled-deflection calibration fixture.

4.5 INSTRUMENTS FOR STATIC RECORDING

In static-stress analysis with photoelastic coatings, photographic records of the full-field and localized fringe patterns are ordinarily necessary. Although on occasion the fringe patterns are traced directly on the coating, photography of the fringes is the more popular method. An ordinary 35-mm camera may be mounted on the tripod for full-field recording or high-magnification recording by photographing through a microscope as shown in Fig. 4.3. The photographic records are used not only to illustrate the full-field fringe patterns but also to pinpoint the locations of critical areas where point-by-point measurements are to be made by using the compensator. The point-by-point data can be recorded in tabular form as suggested by Table 4.1. When large numbers of point-by-point observations are to be made perhaps under several different load levels, data collection can be greatly facilitated by using a digitized compensator with a direct printout (Fig. 4.5).

Fig. 4.6—Controlled-deflection calibrator.

Recording several families of isoclinic fringes can be accomplished by photographing each individual isoclinic pattern on 35-mm color-slide film. The choice of film is not critical because the isoclinics are black. After development, each slide is projected with the images focussed on a white poster board. A complete family of isoclinics is obtained on a single record by tracing the isoclinic lines from each of the slides.

Recording the isochromatic fringe patterns requires careful film selection and close control of the exposure time since the quality of the negative and the colored prints is seriously affected by either overexposure or underexposure. In photographing curved surfaces, care must be taken to insure uniform illumination, and it may be necessary to employ more than one polarized light source. Also, the

Table 4.1—Point-by-Point Data Record Form

Location Point	Fringe Value $F_\epsilon = \dfrac{f_\epsilon}{2 h_c}$	Measurements			Difference of		Individual Stresses	
		Direction ϕ	Normal fringe order N_0	Oblique fringe order N_θ	Principal strains $\epsilon_1 - \epsilon_2 = NF_\epsilon$	Principal stresses $\sigma_1 - \sigma_2 = \dfrac{Nf_\sigma}{2h_c}$	σ_1	σ_2
1								
2								
3								

smallest possible aperature (f-stop) should be used to increase the depth of field.

For black and white photography, exposure time is less critical since such films exhibit a greater latitude over which exposure produces satisfactory results. Monochromatic light is used to improve contrast. Usually this is accomplished by inserting an interference filter with a narrow band width centered at about 5700–5800 Å.

4.6 INSTRUMENTS FOR DYNAMIC RECORDING

The photoelastic-coating method is well suited and often used for dynamic stress analysis, under either repetitive or transient loading. As with any other method of experimental stress analysis, dynamic phenomena too rapid for direct visual observation must be sampled cyclically or recorded for subsequent analysis. Representative techniques, instruments, and accessories to the basic reflection polariscope for dynamic strain recording are described in the following sections.

Stroboscopic Lighting

When the coated test object is subjected to a constant-amplitude cyclic load, the fringe patterns can be observed or recorded by illuminating the coating with a stroboscopic light. By adjusting the repetition rate of the light to the load frequency (or a submultiple thereof), the object, with its fringe pattern, appears static. With slight variations in the lighting frequency, the dynamic event can be made to appear in slow motion.

The stroboscopic lighting technique is widely used with photoelastic coatings for the stress analysis of rotating machinery, such as centrifuges and blowers, and for test parts mounted on electrodynamic shakers. It is also used in the stress analysis of reciprocating and oscillating machinery, in short, wherever the loading is cyclically repetitive. For these purposes the light source of the reflection polariscope is replaced by an appropriate stroboscopic light (Fig. 4.7). Frequency adjustment is controlled either manually or by synchronizing the stroboscopic light with the motion of the test specimen by using a displacement pickup to detect the same point in each cycle. As an example, a high-contrast line can be traced on the shaft of a rotating member, and a photocell monitoring the light reflected from the shaft will provide a synchronizing signal to the stroboscope. A magnetic pickup can also be used for synchronization. Similarly, a specimen in a fatigue-testing machine can actuate

Fig. 4.7—Reflection polariscope equipped with a stroboscopic light.

an electric switch each time it reaches maximum amplitude, and thus provides synchronizing signals to the stroboscope. There are obviously many other ways to generate a synchronizing signal, the choice depending upon the nature of the test specimen, its environment, and the availability of pickups.

Under stroboscopic lighting, the photoelastic measurements are made in the same manner as in static situations; i.e., full fringe orders are determined by direct observation, and fractional orders by either the Tardy or the Babinet-Soleil compensator. Special motion-picture cameras permitting synchronized film advance are used to make a dynamic record of the stroboscopically illuminated fringe patterns. Conventional motion-picture cameras are not suitable for this purpose.

Although the stroboscopic lights currently in use are limited to about 200 Hz, test frequencies much higher than this can be employed since it is sufficient to illuminate every second, third, or even tenth cycle. The maximum frequency that can be observed when using a stroboscopic light is limited by the actual duration of the flash in each cycle. As an example, assume a stroboscopic light with a flash duration of 10 μs is flashing at its maximum frequency of 200 Hz. It is necessary to observe the photoelastic-fringe pattern in a rotating wheel, 6 in. (150 mm) in diameter. The maximum rotational speed (RPM) at which the pattern can be observed without a blur, if a 0.020-in. (0.5-mm) displacement occurs during illumination, can be estimated as:

$$(n\pi D/60)\, T_0 \leqslant 0.020 \text{ in.}$$

$$n \leqslant 6000 \text{ RPM} \tag{4.2}$$

A similar calculation will show that small-amplitude vibrations up to about 5000 Hz can be observed when using commercial stroboscopic lights.

Motion-picture Records[25-29]

For photoelastic-coating studies of nonrecurring transient events (such as explosions, impacts, and crack propagation) motion-picture techniques, generally with high-speed cameras, are necessary. Motion-picture recording of photoelastic-coating fringes is basically the same as still photography through a reflection polariscope. However, some care is necessary to protect the polariscope from heat generated by the high-intensity light source required for illumination.

In general, the following considerations should be taken into account before deciding upon the camera system to use for recording transient fringe patterns: (1) the length of time during which the phenomenon is to be filmed; (2) the direction of motion of the event to be recorded; (3) the intensity and duration of the light source; (4) the camera speed (frames per second), i.e., the total number of frames available in one "shot"; and (5) the synchronizing mechanism for triggering the camera and/or light source at the desired moment and for the desired time. High-speed cameras can perform at framing rates up to 25,000 frames/s. These cameras use rolls of film 100–400 ft long (30–120 m), and run from a fraction of a second to a few minutes, depending upon the framing rate. The total running time includes an acceleration period during which a portion of the film roll is exposed at a lower than prescribed speed. Adequate illumination can usually be obtained with floodlights. Ultrahigh-speed cameras operate in the range from 100,000 to 2,000,000 frames/s. In these cameras the film is stationary, and a high-speed revolving mirror projects the images onto the film through appropriate optics. Since the total number of frames is very limited, very precise synchronization is required between initiation of the camera action and the phenomenon being recorded. Extremely intense light sources are required for use with these cameras because of the short exposure time per frame. Obtaining sufficient light for adequate exposure is further complicated by the light losses in the polariscope (60–70 percent) in reflection from the matte surface of the reflective cement and by the necessity for relatively slow-speed color film. Special electronic flash lamps, which provide the required intensity and display constant illumination for a preset period, are required with these ultrahigh-speed cameras.

For laboratory work with a· limited budget, spark-illumination cameras can be used. With these systems the test specimen is illuminated very briefly (about 1 μs) by a spark obtained from a capacitor discharge. An appropriate optical arrangement projects the image onto the film.

The principal advantage of motion-picture photography of photo-elastic coatings is the full-field nature of the record. Whenever the strain is great enough to produce a response of one or more fringes in the coating, full-field recordings of dynamic fringe patterns can be made with motion-picture techniques. However, since fractional fringe orders cannot be obtained from dynamic records, the method is accurate only along points located on a fringe where a resolution of ±0.02 fringe can be achieved under ideal circumstances. Other areas can be explored by the color-matching technique, but the resolution is limited to ±0.15 fringe at best. Fractional fringe orders can also be read from a photographic record of photoelastic fringes with the aid of a densitometer.

Photoelectric Recording

For recording very low static or dynamic birefringence (i.e., a small fraction of a fringe) or for high-resolution measurements, photocells can be used advantageously. The method is limited, however, to point-by-point measurements. In practice, a photocell is placed behind the analyzer of the polariscope, and an appropriate optical system is arranged to project the image of the point under study onto the photocell grid. Changes in birefringence produce changes in the light intensity, which in turn changes the photoelectric current. The current is conducted to an oscilloscope or oscillograph for observation or recording. A photocell system can provide sensitivities to 0.0001 fringe if necessary, and response times of 1 μs or better.

The light intensity I emerging from the analyzer of a polariscope can be expressed as $I = I_0 \sin^2 \pi N$, where I_0 is the maximum intensity. Examination of this function shows that a number of different values of N give the same intensity. To eliminate this ambiguity, it is necessary to select the coating thickness so that the maximum response of the coating does not exceed 0.5 fringe.

Photocell recording has the following limitations that must be kept in mind when considering applications of this technique: (1) calibration of the entire system (including the coated part) is required for every point at which a measurement is to be made, (2) full-scale measurement is limited to less than 0.5 fringe, and (3) the system is influenced by variations in ambient light intensity.

Television Observation and Recording

In situations where direct observation of the photoelastic coating during the test is not practicable or where the event must be observed or monitored in progress, closed-circuit television can be used very effectively. With the television camera mounted directly behind the analyzer of the polariscope, the photoelastic pattern, either in black and white or in color (preferably the latter), can be observed at one or more remote locations. Such systems have been used for strain measurements on the outer skin of a submerged submarine, inside a pressurized rocket case, and in similar situations.

A remotely controlled drive is used to point the polariscope and its television camera at any area to be observed. Quantitative point-by-point measurements of the fringe order are made with a remotely operated compensator. Records of the fringe patterns are made with a conventional camera from the television receiver screen. Only low-frequency dynamic strains can be recorded with these systems because of the stroboscopic effect of television.

4.7 SUMMARY

The basic instrument for strain measurement with photoelastic coatings is the reflection polariscope. When equipped with appropriate optical systems, cameras, light sources, compensators, and other accessories, it has the following capabilities:

1. Point-by-point or full-field measurement of principal strain directions (isoclinics), magnitudes of maximum shear stress or shear strain (normal-incidence isochromatics), and magnitudes and signs of principal strains (oblique-incidence isochromatics).
2. Strain measurements under either static or dynamic conditions and in either elastic or plastic strain fields.

3. Exploration of strain distributions over very large areas (such as an airplane wing) or in very small, localized areas (such as individual crystals, the root of a notch, etc.)

4. Readily obtained quantitative measurements in industrial applications, with resolutions of ± 2 deg in principal-strain directions and ± 10 microstrain in strain magnitudes ($\pm 10 \times 10^{-6}$ m/m).

5. Adaptability to motion pictures, television, stroboscopic lighting, or photocells to provide direct observation or recording of static or dynamic phenomena.

5

PARAMETERS INFLUENCING
ANALYSIS OF DATA

The analytical procedures developed in Chapter 2 for treating the data obtained from photoelastic coatings are based on an idealized model in which the strain is transmitted to the coating without amplification or attenuation and in which the strain is uniformly distributed through the thickness of the coating. Other simplifying assumptions employed in the development, either explicitly or implicitly, include isothermal test conditions, linear-elastic material behavior, and a static stress state. As the applications of photoelastic coatings are extended to more general and complex problems, several parameters not considered in Chapter 2 may become significant and affect the behavior of the coating. In most instances the effects of these parameters can be quantified, and corrective techniques can be employed in the data reduction to suitably account for the influence of each parameter.

Several different parameters—including coating thickness, reinforcement, curvature, strain gradients, Poisson's ratio mismatch, temperature fields, and wave propagation and vibration (which influence the response of the coating)—are treated in this chapter. The approach here is to show the influence of each parameter and to introduce corrective procedures for minimizing or canceling errors.

5.1 EFFECTS OF COATING THICKNESS

When a photoelastic coating is applied to a specimen, the coating becomes an extension of it. In only certain instances are the strains transmitted to the coating without some modification or distortion. In reality, the coating represents a three-dimensional extension of the specimen; and it is loaded by means of shear and normal tractions at the interface. These tractions must vary so that the displacement fields of the coating and specimen are identical at the interface (as dictated by perfect bonding). Thus, in the most general case, the average strain in the coating does not equal the strain at the interface, a strain gradient exists through the thickness of the coating, and the coating reinforces the specimen.

These three thickness effects tend to vanish as the coating thickness approaches zero. However, coatings of finite thickness, usually 0.020–0.120 in. (0.5–3.0 mm), are required to obtain adequate optical response from the coating for accurately reading fringe orders. The basic question concerns the extent of the error introduced by these thickness effects in the application of the coating to different classes of stress-analysis problems. This question is complex and has been the subject of controversy during the past decade. Certain aspects still must be solved; however, progress has been made in understanding many of the factors contributing to the thickness effects of photoelastic coatings and in providing corrections to account for these effects.

Consider the simple model of a coated specimen discussed in Chapter 2. This model will become more complex in a step-by-step fashion as the individual factors influencing the coating behavior are introduced. Where possible, experimental verifications will be used to justify assumptions or to minimize the complexity of the analysis. The most general theoretical treatment of the effect of coating thickness, which analyzes coating behavior in the presence of very sharp strain gradients, will be introduced late in the development.

Reinforcing Effect[9]

Consider a structural component with a photoelastic coating and subjected to a state of plane stress. A principal element removed from the component is shown in Fig. 5.1. The influence of the portion of the applied load carried by the coating can be determined by equating the forces acting on the composite element to the forces acting on the same element taken from the uncoated structural component. Thus

$$\sigma_1^u h_s dy = \sigma_1^s h_s dy + \sigma_1^c h_c dy \qquad \sigma_2^u h_s dx = \sigma_2^s h_s dx + \sigma_2^c h_c dx \qquad (5.1)$$

where σ_1^u, σ_2^u are the principal stresses in the uncoated component. Again assuming perfect strain transmission and a plane state of stress for both the coating and the specimen, eq. (2.26) applies. Combining eqs. (5.1), (2.2), and (2.26) yields:

$$\epsilon_1^u - \epsilon_2^u = (1/C_1)(\epsilon_1^c - \epsilon_2^c)$$

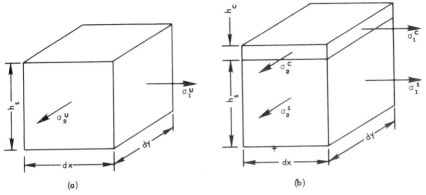

Fig. 5.1—Principal element taken from a plane-stress specimen: (a) with coating, and (b) without coating.

where

$$1/C_1 = 1 + [(h_c E_c)\,(1 + \nu_s)/(h_s E_s)\,(1 + \nu_c)]$$

The term $1/C_1$ represents the correction factor required to convert the strains measured in the coated specimen ($\epsilon_1^c - \epsilon_2^c$) to the strains $\epsilon_1^u - \epsilon_2^u$ that would be developed in the component without a coating. If the term $(1 + \nu_s)/(1 + \nu_c)$ is taken as unity (i.e., neglecting the Poisson's ratio mismatch, which is usually small), then $1/C_1$ is closely approximated by

$$1/C_1 = 1 + (h_c/h_s)\,(E_c/E_s) \qquad (5.2)$$

The correction factor C_1 is shown as a function of h_c/h_s in Fig. 5.2 for a number of common engineering materials with a photoelastic coating having $E_c = 4.2 \times 10^5$ psi (2900 MPa) and $\nu_c = 0.36$. From these results it is evident that the reinforcing effect is quite small when photoelastic coatings are employed with metallic materials subjected to plane stress. For this state of stress the reinforcing effect is significant only in cases where coatings are employed on materials of low elastic modulus or on very thin specimens.

Strain Variations through the Coating Thickness[9, 30]

Thus far only plane-stress applications, for which the in-plane strains ϵ_1^s and ϵ_2^s are constant with respect to position along the normal or z axis, have been considered. If structural components in which ϵ_1^s and ϵ_2^s vary with respect to z (e.g., in bending) are considered, the strains in the coating will also vary through the thickness of the coating. This effect is demonstrated by considering a beam with a photoelastic coating bonded to its upper surface and subjected

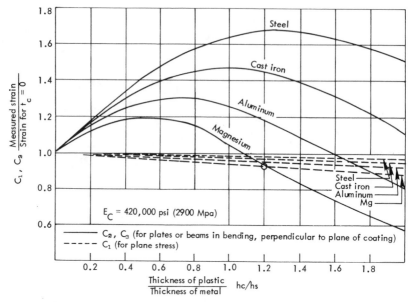

Fig. 5.2—Reinforcing correction factors C_1 and C_2 as a function of h_c/h_s for different specimen materials.

to a constant bending moment M. Removing as a free body an elemental section of the beam (as shown in Fig. 5.3), assuming that plane sections remain plane in bending, indicates that the strains are linearly distributed with respect to z through both the specimen and the coating. Hence

$$\epsilon_1^s = z/\rho \text{ for } -(h_s - A) \leqslant z \leqslant A$$

$$\epsilon_1^c = z/\rho \text{ for } A \leqslant z \leqslant (A + h_c)$$

$$\epsilon_2^s = -\nu_s \, \epsilon_1^s \qquad \epsilon_2^c = -\nu_c \, \epsilon_1^c \tag{5.3}$$

where ρ is the radius of curvature of the deformed beam.

Considering equilibrium of the forces acting on the element along the axis of the beam and noting that the stresses are uniaxial leads to

$$\frac{b}{\rho} \left(E_s \int_{A-h_s}^{A} z\,dz + E_c \int_{A}^{A+h_c} z\,dz \right) = 0$$

This expression provides a relationship for A that positions the neutral axis of the (composite) beam as

$$A = \frac{h_s}{2} \left[\frac{1 - (E_c/E_s)\,(h_c/h_s)^2}{1 + (E_c/E_s)\,(h_c/h_s)} \right] \tag{5.4}$$

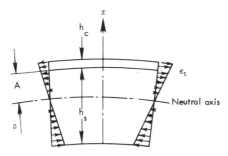

Fig. 5.3—Section of a coated beam subjected to a constant bending moment.

The shift of the neutral axis from the position for the uncoated beam, given by $A = h_s/2$, is due to the reinforcing effect of the coating. Thus this simple example includes both the reinforcement effect and the effect of strain variation through the thickness.

Consider, again, equilibrium of moments about the origin (Fig. 5.3), from which

$$\frac{b}{\rho}\left[E_s \int_{A-h_s}^{A} z^2 dz + E_c \int_{A}^{A+h_c} z^2 dz\right] = M$$

Solution of this equation for the curvature and elimination of A by substituting from eq. (5.4) gives

$$1/\rho = (12M/bE_s h_s^3)\,(1/H) \tag{5.5}$$

where $H = 4(1 + eg^3) - 3(1 - eg^2)^2/(1 + eg)$

$e = E_c/E_s$

$g = h_c/h_s$

Equation (5.5) demonstrates that the reinforcing effect of the coating modifies the beam curvature by the factor $1/H$.

The average value of the principal-strain difference in the coating is obtained from eq. (5.3) as

$$(\epsilon_1^c - \epsilon_2^c)_{\text{ave}} = \frac{1 + \nu_c}{\rho\,h_c}\int_{A}^{A+h_c} z\,dz \tag{5.6}$$

Substituting eqs. (5.4) and (5.5) into eq. (5.6) gives

$$(\epsilon_1^c - \epsilon_2^c)_{\text{ave}} = (6M/H)\,[(1 + \nu_c)/bE_s h_s^2]\,[(1 + g)/(1 + eg)] \tag{5.7}$$

In comparison, the strain difference $\epsilon_1^u - \epsilon_2^u$ at the surface of the same beam (subjected to the same bending moment) without a coating is

$$\epsilon_1^u - \epsilon_2^u = 6M(1 + \nu_c)/bE_s h_s^2 \tag{5.8}$$

Equating eqs. (5.7) and (5.8) with a correction factor $1/C_2$, introduced to account for the difference, leads to

$$\epsilon_1^u - \epsilon_2^u = (1/C_2)(\epsilon_1^c - \epsilon_2^c)_{ave}$$

where

$$\frac{1}{C_2} = \frac{1 + eg}{1 + g}\left[4(1 + eg^3) - \frac{3(1 - eg^2)^2}{1 + eg}\right]\frac{1 + \nu_s}{1 + \nu_c} \tag{5.9}$$

The correction factor $1/C_2$ accounts for the two thickness effects occurring in this example. The first is a reinforcing effect that reduces the beam curvature for a given applied bending moment (and thus the strain at the interface between the coating and the specimen) in comparison to an uncoated beam. The second effect is due to the nonuniform distribution of strain through the thickness of the coating. The optical response of the coating is proportional to the average strain, which in this instance is equal to the strain at the midpoint in the coating. Since the average strain is higher than the interface strain, the effect of strain variation through the thickness of the coating tends to offset the reinforcement of the coating.

In practical applications of photoelastic coatings to beams, h_c is usually much smaller than h_s; the reinforcing effect is small, $H \rightarrow 1.0$, $A \rightarrow (h_s/2)$; and the correction factor can be approximated by

$$1/C_2 \doteq (1 + 4\,eg)/(1 + g) \qquad \text{for } g \leqslant 0.1 \tag{5.10}$$

when the Poisson's ratio mismatch parameter $(1 + \nu_s)/(1 + \nu_c)$ is nearly unity.

Several other examples of structural members showing similar effects of the coating thickness have been studied by Zandman and his associates.[9,30] These include flexure of flat plates, torsion of circular shafts, and pressure loading of a thick-walled cylindrical pressure vessel. In all cases, correction factors $1/C_n$ are derived so that the surface strains on the uncoated object can be calculated from the photoelastic measurement of the coating strains according to

$$\epsilon_1^u - \epsilon_2^u = (1/C_n)(\epsilon_1^c - \epsilon_2^c)_{ave}$$

For the bending of a flat plate of thickness h_s with moments M_1

and M_2, the correction factor is

$$1/C_3 = \frac{1 + emg}{1 + g} \left[4(1 + emg^3) - \frac{3(1 - emg^2)^2}{1 + emg} \right] \quad (5.11)$$

where $m = (1 - \nu_s^2)/(1 - \nu_c^2)$. Note that for $\nu_s = \nu_c$ this equation becomes identical to eq. (5.9), and the numerical values of C_2 and C_3 computed for most engineering materials are the same (see Fig. 5.2). For the torsion of a hollow circular shaft with an inside radius a, outside radius b, and a radius to the surface of the coating $C = b + h_c$, the correction factor $1/C_4$ is given by

$$1/C_4 = \frac{2}{1 + (c/b)} \left\{ 1 + \frac{G_c[(c/b)^4 - 1]}{G_s[1 - (a/b)^4]} \right\} \quad (5.12)$$

where G is the modulus of elasticity in shear. Finally, the correction factor $1/C_5$, for a thick-walled pressure vessel with closed ends and subjected to internal pressure, is given by

$$C_5 = \left[\frac{1 + \nu_c}{1 + \nu_s} \right] \left[\frac{2\left(1 - 2\nu_c + \frac{c}{b}\right)(1 - \nu_s)}{\frac{1 + \nu_c}{1 + \nu_s}\left(1 - 2\nu_c + \frac{c^2}{b^2}\right) + em\left(\frac{c^2 - b^2}{b^2 - a^2}\right)\left(1 - 2\nu_s + \frac{a^2}{b^2}\right)} - \frac{1 - 2\nu_s}{1 + em\frac{c^2 - b^2}{b^2 - a^2}} \right] \quad (5.13)$$

Numerical values of the correction factors C_3, C_4, and C_5 are given in Figs. 5.4, 5.5, and 5.6.

5.2 EFFECT OF POISSON'S RATIO MISMATCH[31]

In plane-stress problems the errors arising from coating-thickness effects (i.e., reinforcement, strain gradients, and curvature) are usually negligible. However, in almost all cases a mismatch in Poisson's ratio occurs, with ν_c usually greater than ν_s. This difference in Poisson's contraction produces a distortion of the displacement field through the thickness of the coating, which is particularly pronounced at the boundaries (Fig. 5.7).

The strain field near the boundary is not uniformly distributed through the thickness of the coating. To examine this effect, assume that the strains ϵ_1 tangent to the boundary are transmitted without loss or amplification so that $\epsilon_1^s = \epsilon_1^c$. This implies that the distortion

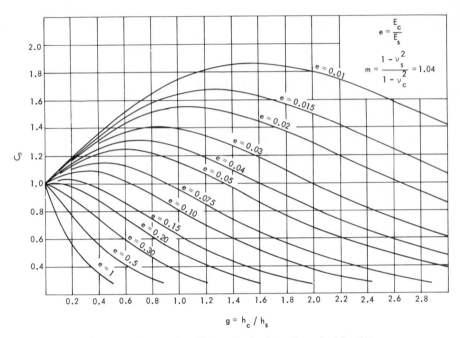

Fig. 5.4—Correction factor C_3 for bending of wide plates.

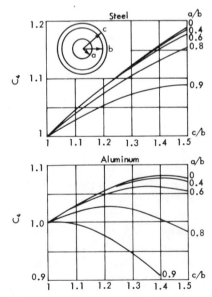

Fig. 5.5—Correction factor C_4 for torsion of circular shafts.

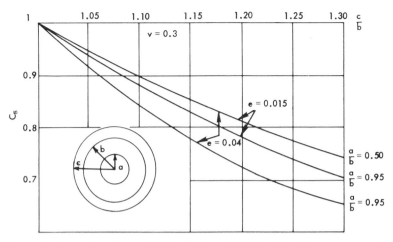

Fig. 5.6—Correction factor C_5 for thick-walled closed-end pressure vessels.

of the strain field occurs only in the transverse direction due to the difference in Poisson's ratio.

At the interface the transverse strain in the coating ϵ_2^c is controlled by the specimen strain, and

$$\epsilon_2^c = \epsilon_2^s = -\nu_s \epsilon_1^s \qquad (5.14)$$

However, at the free surface of the coating the transverse strain is determined solely by the Poisson contraction of the coating material, with

$$\epsilon_2^c = -\nu_c \epsilon_1^c = -\nu_c \epsilon_1^s \qquad (5.15)$$

The average value of ϵ_2^c through the thickness of the coating is obviously bounded by these two limiting values. Thus from eq. (2.8) and the strains described in eqs. (5.14) and (5.15), the fringe-order

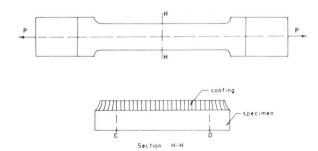

Fig. 5.7—Distortion of a displacement field through the thickness of the coating, due to Poisson's ratio mismatch.

response of the coating at a free boundary is bounded by

$$\epsilon_1^s (1 + \nu_s)/F_\epsilon < N < \epsilon_1^s (1 + \nu_c)/F_\epsilon$$

where $F_\epsilon = f_\epsilon/2h_c$.

It can be seen from this inequality that the magnitude of the strain distortion is determined by the Poisson's ratio mismatch parameter $(1 + \nu_c)/(1 + \nu_s)$. Assuming a fixed value of $\nu_c = 0.36$ and varying ν_s between 0.0 and 0.5, the mismatch parameter varies from 1.36 to 0.90. Experiments with the tensile specimens fabricated from glass-fiber-reinforced plastics that were designed to give ν_s from 0.10 to 0.35 (Fig. 5.8) demonstrate that the fringe orders on the boundary and at interior regions are given accurately by

$$N = \epsilon_1^s (1 + \nu_c)/F_\epsilon \qquad \text{on the boundary}$$

$$N = \epsilon_1^s (1 + \nu_s)/F_\epsilon \qquad \text{in the interior region}$$

A transition zone, where the optical response of the coating is changing, exists near the boundary where the relationship between

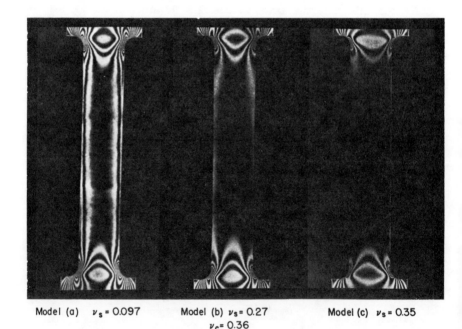

Model (a) $\nu_s = 0.097$ Model (b) $\nu_s = 0.27$ Model (c) $\nu_s = 0.35$
$\nu_c = 0.36$

Fig. 5.8—Effect of Poisson's ratio mismatch on response of photoelastic coatings bonded to glass-fiber-reinforced plastics: Model (a) $\nu_s = 0.097$; Model (b) $\nu_s = 0.27$; $\nu_c = 0.36$; Model (c) $\nu_s = 0.35$.

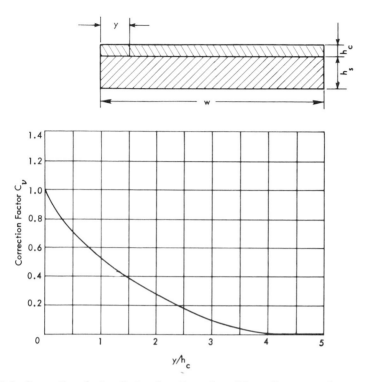

Fig. 5.9—Correction factor C_ν is a function of position y/h_c across the transition zone.

the observed fringe order and the strain in a tension specimen can be expressed as

$$N = [1 + \nu_s + C_\nu \ (\nu_c - \nu_s)] \ (\epsilon_1^s / F_\epsilon) \qquad (5.16)$$

where C_ν is a correction factor accounting for the mismatch effect.

Other experiments by Dally and Alfirevich,[31] conducted with tension specimens having a large mismatch parameter $(1 + \nu_c)/(1 + \nu_s) = 1.24$ and with coatings ranging in thickness from 0.022 to 0.130 in. (0.56 to 3.3 mm), indicate that the width of this transition zone is about four times the thickness of the coating. The value of C_ν decreases from 1.0 at the boundary to zero at a distance of $4\,h_c$ from the boundary, as shown in Fig. 5.9. For metallic-specimen materials, $\nu_c - \nu_s$ is usually less than 0.06, which is relatively small in comparison to $1 + \nu_s = 1.36$; and the Poisson's ratio mismatch can often be neglected.

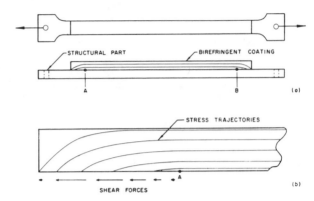

Fig. 5.10—Transfer of forces from specimen to coating.

5.3 INFLUENCE OF STRAIN GRADIENTS[32]

Plane Stress

The specimen and the coating are assumed to have matched values of Poisson's ratio, the mismatch parameter $(1 + \nu_c)/(1 + \nu_s) = 1.05$ is small enough to be neglected in many applications. Consider first a flat tension specimen with a coating, since this represents the simplest possible case of plane stress. The deformation of the coating is introduced by the action of shear forces near the ends of the coating, as illustrated in Fig. 5.10. The shear forces are highly concentrated near the edge of the coating but diminish to zero within a few coating thicknesses from the edge, at points A and B. Within zone AB both the specimen and the coating are in a state of plane stress. The photoelastic coating behaves as an independent member subjected to the same axial displacement as the specimen between A and B. This implies that the strain field in the coating depends only on the bond near the edges since no shear tractions exist between the specimen and the coating in region AB. In this region the specimen serves only as a support for the coating and as a front-surface mirror to reflect the light.

Next, introduce a variety of geometric discontinuities in region AB, as shown in Fig. 5.11. The specimen and the coating over this region are subjected to the same end displacements; and the loads necessary to create these displacements are completely transferred in the border zones outside A and B. Since the coating and specimen are of the same geometry and are given the same boundary deformations (over region AB), the displacement and strain fields in both the coating and specimen are identical. Therefore, a bond between the coating and specimen is not necessary in region AB. However, the

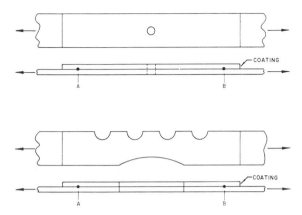

Fig. 5.11—Examples of discontinuities in plane-stress problems.

bond in the shear zones outside A and B must be sufficient to transmit the required shear tractions at the edges of the coating.

The correspondence between fringe patterns obtained in bonded and unbonded coatings was demonstrated by Post and Zandman[32] by means of the multiply notched specimen shown in Fig. 5.12. The specimen was made from 2024-T4 aluminum alloy, and machined to a thickness of 0.178 in. (4.52 mm). Birefringent coatings 0.750 in. (19.1 mm) thick with $\nu_c = 0.36$ were symmetrically located on each side of the specimen. An array of semicircular notches with 1/8 in. (3.2 mm) radius and 3/8 in. (9.5 mm) between centers was introduced in the composite member to represent geometric discontinuities. Before bonding, a silicone grease was applied to one portion of the polished metal surface to ensure that part of the surface remained unbonded. The fringe pattern obtained with this specimen is shown in Fig. 5.13. The high-order fringes permit immediate comparison of patterns in the bonded zone (left of scribed line) and the unbonded

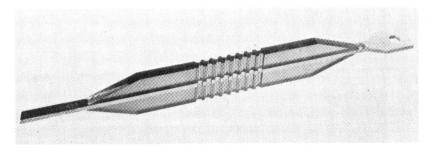

Fig. 5.12—Multiply notched specimen with thick birefringent coatings.

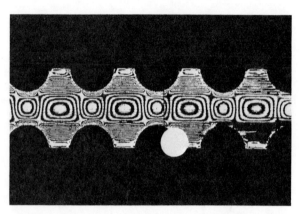

Fig. 5.13—Identical fringe patterns exhibited by bonded (left) and unbonded
(right) sections of the multiply notched specimens.

zone (right of line). The patterns in both zones are identical, and the
case for zero shear tractions at the coating interface is verified.

This is an important finding since it demonstrates that in-plane
strain gradients do not affect the accuracy of the coating indication
in any analysis of a structure subjected to plane stress. There remains
an error due to a mismatch in Poisson's ratio, but in the case of metal
test specimens the error is less than 5 percent since the mismatch
$(\nu_c - \nu_s)/(1 + \nu_s)$ is usually less than 0.05. Moreover, on the boundary, where the effect of the mismatch is greatest, the data can be corrected by using the correction factor C_ν and eq. (5.16) to eliminate
the error.

General Surface Strain

In certain cases the strain distribution in the coating will differ
from that on the specimen surface, even with the same displacements
applied at the ends of both components. An example of this condition is the tension member with a subsurface transverse hole depicted
in Fig. 5.14. The displacements at sections A and B are identical in
the coating and the structure. In this instance, however, there is a
local concentration of strain near the central hole which will not be
transferred unless the coating is bonded to the tension member. Ac-
cordingly, the deformation of the coating in this region is developed
to some degree by the action of local shear tractions at the interface.
The shear tractions produce a state of strain in the coating which is
not independent of the (z) coordinate, and the birefringence developed varies through the thickness of the coating.

Duffy et al.[33-35] have considered this problem of the thickness
effect by treating the coating as a three-dimensional elastic body and

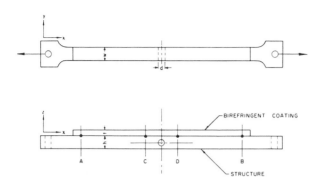

Fig. 5.14—Example of a general surface strain (shear tractions are not zero in zone CD).

prescribing a displacement field at the interface between the specimen and the coating. Solutions were obtained for the strain distribution through the thickness of the coating. Considering the simplest of the three cases treated by Duffy, the displacement field at the interface (u_0, v_0, w_0) is prescribed by a Fourier series:

$$u_0 = \frac{A_0}{2} + \sum_{n=1}^{\infty} A_n \cos p_n x + \sum_{n=1}^{\infty} B_n \sin p_n x$$

$$w_0 = \frac{C_0}{2} + \sum_{n=1}^{\infty} C_n \cos p_n x + \sum_{n=1}^{\infty} D_n \sin p_n x$$

$$v_0 = 0 \qquad\qquad\qquad\qquad (5.17)$$

where the coordinate system is defined in Fig. 5.15, and

A_0, C_0 and A_n, B_n, C_n, D_n = Fourier coefficients necessary to describe an arbitrary displacement field

$p_n = 2\pi/\lambda_n$, the wave number

λ_n = wavelength of the displacement field

The coating was treated as a three-dimensional elastic body, and an exact solution for the displacement field, $u = u(x, z)$, $v = 0$, $w = w(x, z)$, was obtained. The solution for $u(x, z)$ is shown below as expressed by the first term in the series expansion:

$$u(x, z) = (A \cos px + B \sin px) \left[(\Gamma_1 - \Gamma_3 pz) e^{pz} + (\Gamma_2 - \Gamma_4 pz) e^{-pz} \right]$$

$$+ (-C \sin px + D \cos px)$$

$$\times \left[(\Gamma_1^* - \Gamma_3^* pz) e^{pz} - (\Gamma_2^* - \Gamma_4^* pz) e^{-pz} - (3 - 4\nu)(\Gamma_3^* e^{pz} + \Gamma_4^* pz) e^{-pz} \right]$$

$$(5.18)$$

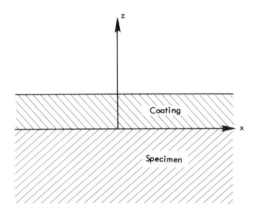

Fig. 5.15—Coordinate system.

where the Γ's are constants of integration, determined to satisfy the boundary conditions (see Refs. 33 and 36 for evaluation). The leading terms in eq. (5.18), containing coefficients A and B, represent the contribution due to the in-plane surface displacement u_0; and the following terms, containing C and D, represent the contribution due to the out-of-plane displacement (w_0), which can produce curvature of the interface.

The average strain difference through the thickness of the coating is

$$(\epsilon_1 - \epsilon_2)_{\text{ave}} = \frac{1}{h} \int_0^h \frac{\partial u}{\partial x} dz = \frac{N f_\epsilon}{2h} \qquad (5.19)$$

since $\epsilon_2 = 0$ as a result of $v = 0$. The expected fringe order is then given by

$$N = \frac{2}{f_\epsilon} \int_0^h \frac{\partial u}{\partial x} dz \qquad (5.20)$$

Substituting eq. (5.18) into eq. (5.20) and integrating gives

$$N = (2/f_\epsilon) \, [Fh(\partial u_0/\partial x) + G \, w_0] \qquad (5.21)$$

where $\partial u_0/\partial x$ = extensional strain at the interface
           w_0 = displacement of the interface normal to its plane

The coefficients F and G in eq. (5.21) are correction factors defined by

$$F = \frac{1}{h} \int_0^h \left[\left(\frac{\partial u}{\partial x} \right)_1 \Big/ \left(\frac{\partial u}{\partial x} \right)_0 \right] dz$$

$$G = \int_0^h \left[\left(\frac{\partial u}{\partial x} \right)_2 \Big/ w_0 \right] dz \qquad (5.22)$$

where $(\partial u/\partial x)_1$ = strain in the coating due to the longitudinal strain
          at the interface

   $(\partial u/\partial x)_2$ = strain in the coating due to the curvature of the
          interface

Substituting eq. (5.18) into eq. (5.22) and integrating yields the
following expressions for the correction factors:

$$F = \frac{2\nu ph \cosh ph - 2(1 - 2\nu) \sinh ph + (1 - 2\nu)(2ph + \sinh 2ph)}{ph\,[(ph)^2 + (3 - 4\nu) \cosh^2 ph + (1 - 2\nu)^2]}$$

$$G = \frac{4\nu(1 - \nu)(1 - \cosh ph) - 2\nu\,ph \sinh ph + (ph)^2 + \sinh^2 ph}{(ph)^2 + (3 - 4\nu) \cosh^2 ph + (1 - 2\nu)^2}$$

$$(5.23)$$

These factors are not particularly sensitive to Poisson's ratio;
therefore, the results are presented in Fig. 5.16 for a value of ν_c =
0.35, which closely matches that of most polymeric materials used as
photoelastic coatings. Inspection of Fig. 5.16 shows that $F \rightarrow 1$ and

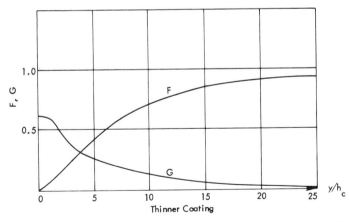

Fig. 5.16—Correction factors F and G as a function of the wavelength-to-
thickness ratio λ/h.

$G \rightarrow 0$ for very thin coatings for which $(h_c/\lambda) \rightarrow 0$. The expected fringe order N, given by eq. (5.21), becomes $(2h_c\epsilon_0/f_\epsilon)$, indicating perfect strain transmission from the specimen to the coating. In the more general case, h_c/λ will be finite; and $F < 1.0$, $G > 0$, and some error will result due to strain variations through the thickness of the coating.

Theoretical examples of displacement and strain fields such as $u_0 = \cos (px)$, $w_0 = v_0 = 0$ have been analyzed by Duffy;[33] and the results show the existence of very large errors. In this case, with $h_c/\lambda = 0.1$, the average strain in the coating is about 65 percent of the specimen surface strain, indicating an error of 35 percent. However, the existence of such large errors in practical applications in static elastic tests, where the experiments are conducted with reasonably thin coatings, has not been demonstrated.

A more practical example of the magnitudes of the errors produced by strain-gradient effects in photoelastic coatings was described by Duffy and Mylonas,[35] who considered the tension specimen with a subsurface transverse hole illustrated in Fig. 5.14. In this case the bar was 2.496 in. (63.40 mm) wide, and the hole diameter was 1.668 in. (42.37 mm), giving $D/W = 0.668$. The surface-strain distribution $(\epsilon_{xx})_0$ is given in Fig. 5.17, where x is measured from the center line of the transverse hole. Based upon this strain distribution, correction factors for strain gradient F and curvature G were computed as a function of position x for a coating 0.125 in. (3.18 mm) thick. The two corrections are essentially opposite in sign, and thus the net correction factor is considerably smaller than the individual values. In this case the ratio of the net correction factor to $(\epsilon_{xx})_0$ at the position of maximum strain is about 3 percent, although it is higher at $x/R = 0$ where $(\epsilon_{xx})_0$ is a minimum.

The theoretical studies using sinusoidal displacement fields for u_0, v_0, and w_0 indicate substantial distortion of the average strain field transmitted to the coating. However, it is difficult to confirm these large errors in the laboratory because strain fields with high-order strain gradients (as implied by sinusoidal displacement fields) cannot readily be produced. Duffy also expressed the specimen-surface displacement fields as polynomials in power series expansions. He showed that constant surface strains or strains that vary linearly in x do not create a variation in strain through the thickness of the coating. Since most elastic strain fields encountered in actual practice contain significant constant and linear terms, the relative errors produced by thickness effects associated with the higher order terms in the strain field are appreciably reduced.

The effects of high-order strain gradients can be demonstrated in

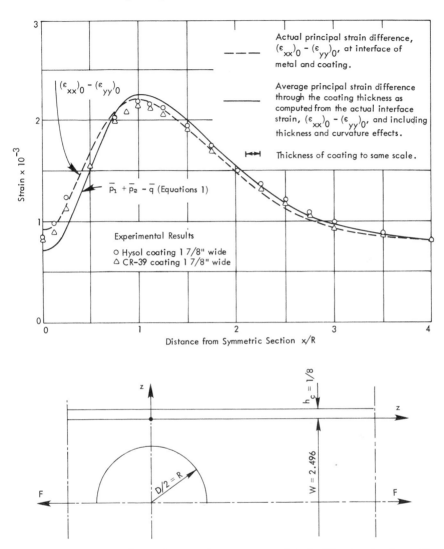

Fig. 5.17—Actual strains and correction terms associated with a tension bar with a subsurface transverse hole [$h = 0.125$ in. (3.175 mm), $D = 1.668$ in. (42.367 mm), and $w = 2.496$ in. (63.398 mm)].

the laboratory by plastically deforming mild steel until a well-defined system of Lueder's lines is evident. These lines represent very narrow bands across which extremely large and permanent shear strains have occurred. An equation expressing the strains across one of these slip bands would require many significant higher order terms since the field is essentially discontinuous across each band.[32]

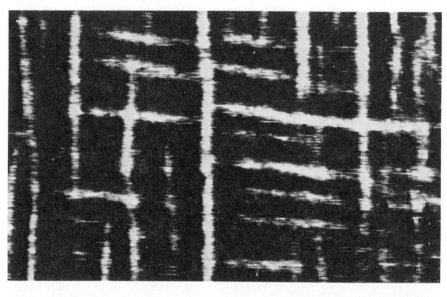

Fig. 5.18—Fringe patterns associated with plastic deformation in mild steel

The photoelastic response of the coating bonded to a mild-steel specimen (shown in Fig. 5.18) indicates the presence of the Lueder's lines. Birefringence measurements were taken along a particular line with a microscope-type polariscope fitted with a compensator. The thickness of the coating was reduced in successive operations to obtain the data shown in Fig. 5.19. The fringe order in regions between the slip bands has decreased in proportion to the reduction in coating thickness, whereas the birefringence spikes are not markedly influenced by coating thickness. Since the slip band is on the order of 1 μin. (25nm) wide, the ratio h_c/λ for the thinnest coating is on the order of 10^4, and huge strain distortions occur through the thickness of the coating. Thus the birefringence at local disturbances is not proportional to coating thickness.

This last example emphasizes the importance of local shear tractions in transmitting strains to the coating. When these shear tractions are required to maintain displacement continuity between the structure and the coating, it is important to minimize the thickness of the coating. For improved accuracy the coating thickness should be small compared to the span over which shear tractions occur.

Finally, in plane-stress problems for which the Poisson's ratio mismatch is small, no significant shear tractions develop between the coating and the specimen, and no inaccuracies occur due to strain gradients through the coating.

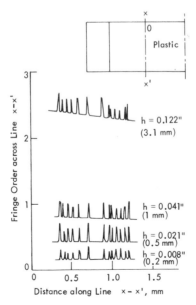

Fig. 5.19—Fringe order as a function of position across a plastically deformed specimen, h = coating thickness.

5.4 INFLUENCE OF TEMPERATURE[37]

Photoelastic coatings can be utilized to measure strains in structures subjected to changing temperatures in the range of -50 to $+400°F$; however, corrective procedures are often necessary to account for the effects of temperature changes. Ideally, a coating perfectly adapted for use in thermal-stress analysis would exhibit the following three special characteristics: (1) thermal conductivity equal to that of the structure, (2) coefficient of thermal expansion identical to that of the structure, and (3) material-fringe value (in terms of strain) invariant with temperature.

Because polymeric coating materials exhibit relatively low thermal conductivity in comparison with metallic materials, the presence of the coating affects the temperature field in transient thermal-stress problems. In these specific problems the coating cannot be employed directly on the prototype structure, and it is necessary to use a model fabricated from a polymer to closely match the thermal conductivity and the thermal coefficient of expansion of the coating. With such coated models the temperature field is simulated, and results for the thermal-strain field are converted by means of scaling relationships to obtain prototype strains.

Photoelastic coatings can be used directly on prototypes in two different types of thermal-stress applications. The first application involves bodies where temperature gradients exist along the surface but where the material is highly insulating with negligible heat conduction. The second application involves bodies with temperatures on two or more surfaces maintained at constant but different levels. In applications of the latter type it may be necessary to deviate from the actual operating temperature of the system to accommodate the useful temperature range of the coating. It is essential that the simulated temperature distribution correspond to the actual distribution. Thus the temperature difference between any two points must remain in constant proportion for simulated and actual conditions. Since thermal stresses are linearly related to temperature differences, results obtained with simulated temperature differences $(\Delta T)_s$ must be scaled to actual temperature differences $(\Delta T)_a$ by $\sigma_a = (\Delta T_a / \Delta T_s) \sigma_s$, where subscripts a and s represent actual and simulated conditions.

For cases where temperature is uniform but markedly different from room temperature, the fringe-order readings must be corrected, taking into account variations of the strain-sensitivity K and of the mismatch of the coefficient of thermal expansion $\alpha_c - \alpha_s$. Typical variations of these properties vs. temperature are shown in Fig. 5.20 for an epoxy coating. Further details on the application of photoelastic coatings in thermal fields are presented in Refs. 37 and 38.

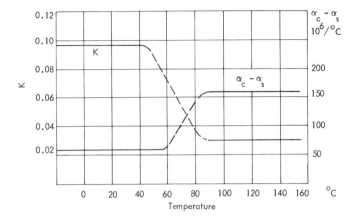

Fig. 5.20—Influence of temperature on K and $\alpha_c - \alpha_s$ for a typical epoxy coating material.

5.5 INFLUENCE OF DYNAMIC STATES OF STRESS[38-44]

The use of photoelastic coatings to study dynamic states of stress requires that the coating have a sufficiently high-frequency response to follow the changes that occur in the stress field with time. To determine the frequency response of a coating, consider an experiment conducted by Duffy and Lee[39] on an aluminum bar with a photoelastic strip bonded to its surface, as illustrated in Fig. 5.21. A stress wave is propagating along the bar with a velocity, $c_L = 2 \times 10^5$ in./s $(5 \times 10^3$ m/s). Because the coating is bonded to the bar, the stress wave traveling along the bar produces interface displacements in both the axial and transverse directions. These displacements in turn excite two stress waves in the coating with velocities $c_1 = 0.8 \times 10^5$ in./s $(2 \times 10^3$ m/s) and $c_2 = 0.4 \times 10^5$ in./s $(1 \times 10^3$ m/s). The higher velocity wave in the coating is a dilatational wave, and the lower velocity wave is a shear wave.[40] It can be seen from Fig. 5.21 that large variations in strain exist through the thickness of the coating and that the observed fringe response through the thickness of the coating lags (with time) the stress in the bar.

Duffy[39] has shown these waves propagating in a birefringent strip bonded to the pressure bar and has exhibited them photoelastically by using transmitted light. These dynamic fringe patterns obtained from the strip (Fig. 5.21) dramatically illustrate the two waves that are excited in the coating. The bonded strip can be used as a very effective means to study dynamic stresses created by wave propagation. As the interface between the strip and the bar can be observed directly, there is no lag time and the question of frequency response does not arise.

Returning to the use of coatings in measuring dynamic stresses,

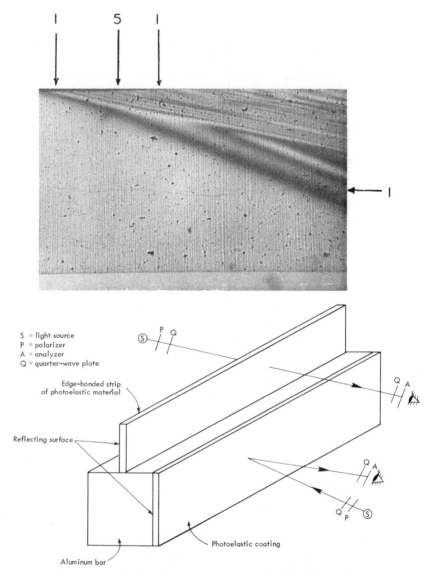

Fig. 5.21—Wave propagation in a pressure bar and in a coating.

the wave propagation across the coating thickness h_c occurs with a maximum transit time $t = h/c_2$. As the stress wave reflects from the free boundary, it is not realistic to assume that the coating responds to the dynamic surface strains in a single transit time; however, after ten transits it is reasonable to expect that the coating has responded. Then the period of response T_R can be approximated as $T_R = 10h_c/c_2$, and the frequency response f_r is given by

$$f_r = 1/T_R = c_2/10h_c \qquad (5.24)$$

From this result it appears that the frequency response of the coating increases as the coating thickness is decreased. For a typical coating 0.1 in. (2.5 mm) thick, it appears that the frequency response is of the order of 40 kHz. This value is sufficiently high that photoelastic coatings can be employed in most dynamic-stress-analysis problems except those involving wave propagation. For the latter conditions, the strip technique is preferred. Other dynamic stress analyses using both birefringent strips and coatings are described in Refs. 27, 29, 41–43.

When employing the coating to measure dynamic strains in vibrating members, care must be exercised so that the coating does not significantly alter the vibrating system. The photoelastic coating increases the energy dissipated in a vibrating system as a result of hysteresis losses in the coating under cyclic straining. If the system is characterized by low damping energy and is being excited at or near resonance, the photoelastic coating will markedly affect the system by reducing the resonance-amplification factor. In these cases, the results obtained with the coatings should be verified with strain gages. In other cases where the inherent damping in the structure is large and the exciting forces are not near the natural frequency of the structure, the additional damping introduced by the coating is not significant; and the results obtained with stroboscopic reflected light are accurate and meaningful. The effect of centrifugal forces such as encountered in rotating parts coated with photoelastic coating were studied by Nickola.[30]

5.6 SUMMARY

This chapter extends the analytical procedures developed in Chapter 2 to include the influences of several parameters that affect the response of the coating. The influence of coating thickness is considered first, with correction factors developed to account for reinforcement and for strain variations through the thickness of the coating that are due to bending, torsion, and pressure. The influence of Poisson's ratio mismatch between the coating and the structure is investigated and shown to be very small in most cases. A correction factor is developed to adjust the results for this effect when the mismatch is significant.

The influence of strain gradient is first considered in a plane-stress case, for which it is shown that strain transmission is essentially accomplished with zero shear stresses and that strain gradients do not affect the behavior of the coating. The influence of strain gradients in a general state of surface strain is also introduced. Realistic examples are discussed to show that the magnitudes of the errors involved in an elastic-stress-analysis problem are usually quite small, while the errors in an analysis of the plastic deformation across a slip line (Lueder's lines) are quite large. Discussion on the use of photoelastic coatings in thermal fields shows that the coatings can be employed in elevated temperature testing if the temperature of the surface is uniform. Corrections are required to adjust for the effect of the difference in coefficient of thermal expansion between the coating and the specimen. An extensive description of the thermal effect is available in Ref. 37.

The behavior of the photoelastic coating under dynamic stress conditions is described. The strip method of photoelasticity is introduced as a means of treating wave-propagation problems. The frequency response of a typical coating is estimated at 40 kHz based on a wave-propagation argument. Thus the coatings are adequate for vibration analyses and other transient problems when damping introduced by the coating is not a major factor.

6

INDUSTRIAL AND RESEARCH
APPLICATIONS

This chapter demonstrates the application of photoelastic coatings to a variety of industrial problems. Unfortunately, many examples must be omitted because of space limitations. The chapter is organized by industry type, with selected applications illustrating certain unique aspects of the coating technique. The following aspects of the photoelastic-coating technique are discussed or illustrated: dynamic-stress analysis, buckling, elastic and elastoplastic analysis, anisotropic materials, thermal stresses, detection of cracks, onset of yielding, rotating bodies, large and small structures, model analysis, and wave propagation.

6.1 AEROSPACE[7,21,45,46]

Aircraft

Many aircraft have been stress analyzed with photoelastic coatings under both static and flight conditions. The method is very well suited for such applications since it provides full-field data on prototype structures. A few examples of its use on aircraft are briefly described in the following sections.

Airplane Wing and Window Frame. The wings and fuel access doors of the Lockheed C-141 military jet transport have been analyzed under compression loads. The analysis of the stress distributions included both the elastic and elastoplastic ranges of deformation. In addition, one requirement was to establish the load that initiated elastic buckling. Buckling was observed during the loading of the wing while recording the photoelastic pattern with a motion-picture camera. The onset of buckling was evident by the sudden appearance of asymmetric isochromatics of high fringe order on one section of the wing. The initiation of the buckling was detected photoelastically before any nonlinearities were observed in the displacement gages or on the load record from the testing machine.

Plate C.2 shows the photoelastic coating pattern on a section of wing under 90,000 lb (20 KN) load. Note that the fuel access door, designed as a nonload-carrying member, exhibits higher stresses than the adjacent area of the wing. Strain-gage rosettes are located at points a and b to provide a second experimental measurement of the strain to verify the accuracy of the results from the coating. Plate C.3 reveals the existence of permanent deformations in the reinforcing ribs of the fuel access door following removal of the load. In this view (the reverse side of the door) the residual isochromatics under no-load conditions testify to elastoplastic buckling of the ribs. Because of the reinforcing effect, care must be taken in selecting the coating thickness for application to thin skins that are subject to buckling. Although the elastic-stress distribution is not usually affected by the reinforcing effect of the coating, the critical load at which buckling starts may be.

Several tests have been performed on the window frame of the Douglas DC-8 jet passenger transport. Plate C.4 shows the isochromatic pattern in the window frame when it was subjected to 98 percent of the maximum load. Note the stress concentrations around the holes.

Flight Tests. In addition to many applications in the ground testing of aircraft, photoelastic coatings have been used successfully in certain in-flight tests. Since the method is optical, a degree of improvization may be required for observing and recording the photoelastic patterns. Wings have been analyzed in flight by using one of the following procedures: (1) during night flights the coating is observed from the cabin through the window, while using a reflection polariscope equipped with a very high-intensity light source, and (2) the coating on the wing is covered directly with a circular polarizer and viewed through the cabin window with daylight illumination.

There are some difficulties with these procedures. The coating must be well bonded to the wing to avoid being torn off by aerodynamic forces during flight. The same requirement applies to a circular polarizer if it is bonded over the coating. In addition, photography of the fringe patterns is difficult in daylight. A lens with long focal length must be used because the pattern appears in a dark area surrounded by very bright light. No problems are encountered, however, with visual observations during day flights. The fringe orders are observed under poorly defined oblique-incidence conditions and corrections are required to obtain relatively accurate quantitative data.

Tests of this type have been very fruitful, often revealing unsuspected sources of trouble. For example, it has been found in certain instances that taxiing on the ground was the most damaging of the load conditions that produced fatigue cracks in the vicinity of the fuel access doors. The loading conditions that were investigated included fueling; taxiing on smooth and rough runways; full-power engine operation with the aircraft stationary; various takeoff, flight, and landing conditions; and actuation of the landing gear.

The aircraft window itself was analyzed in flight, with a coating applied to the outside of the window and with the outer surface of the coating painted with reflective aluminum paint. The aircraft was flown under different cabin pressure conditions, and the analysis was performed from inside the cabin, as shown in Fig. 6.1. The birefringence in the window glazing is low because it is made of low-sensitivity material and is subjected primarily to out-of-plane bending, which produces a negligible birefringence through the thickness. Thus the photoelastic readings quite accurately revealed the strain on the outside surface of the window. These photoelastic measurements must be corrected for bending and reinforcing effects, as described in Chapter 5.

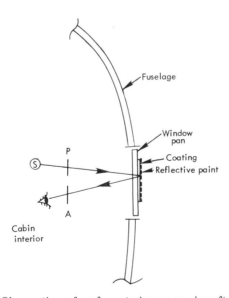

Fig. 6.1—Observation of surface strains on an aircraft window.

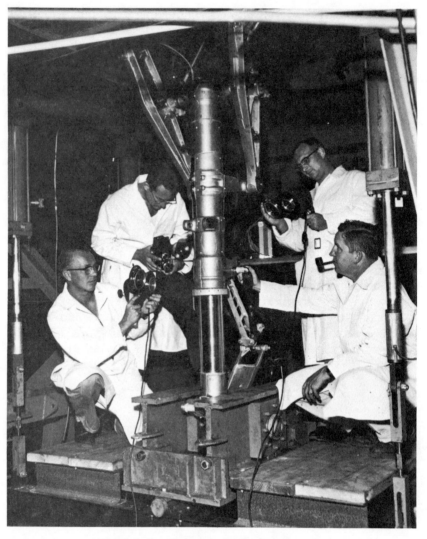

Fig. 6.2—Landing gear of a B-58 under test.

Landing Gears.   The landing gears for nearly all modern aircraft have
been stress analyzed by covering the entire gear surface with photoelastic coating.   These include the landing gears of the Boeing 707,
720, 727, and 747; the Lockheed Jetstar, C-131 and C5-A; the Douglas DC-8, DC-9, and DC-10; the Convair B-58 and B-1 bombers; the
Concorde; etc.

Landing gears are fabricated from forged and machined high-
strength steel.   The gear is a complex assembly of parts subjected to

various static and shock loadings. Occasionally, certain parts are exposed to as many as six different loading conditions. Because the landing gear is used only twice during a flight and represents dead weight the remainder of the time, any weight reduction is of great benefit. At the same time, safety is obviously of paramount importance; large safety factors must be employed unless the stress distribution is accurately known for all significant modes of loading.

Figure 6.2 is a general view of the Convair, General Dynamics Division, B-58 supersonic bomber landing gear under test. Technicians are recording isochromatic and isoclinic fringe patterns with portable reflection polariscopes. The isochromatic fringe pattern in a portion of the B-58 landing gear is shown in Fig. 6.3. This portion of the gear is subjected to combined torsion and bending. Note that the isochromatic fringes are continuous through the junction of two adjacent photoelastic sheets. This will always occur when there is continuity in the coating (the gap between the two adjacent sheets is filled with cement to present a continuous plastic surface).

A general view of the Boeing 747 main landing gear is given in Fig. 6.4.[45] The landing gear for this test was cast and machined from ModelTech (trademark, Photolastic, Inc.) material (an easily castable aluminum-filled epoxy with an elastic modulus of 1×10^6 psi, or 7000 MPa). The full-scale 747 gear model made of this material was tested under scaled-down loads and analyzed with photoelastic coatings and strain gages to arrive at the optimum shape.[45] In some regions material was removed by machining; in others material was added by bonding or casting ModelTech material onto the gear. After several such alterations of shape, the metal prototype was fabricated and retested with coatings and strain gages. Plate C.5 shows the isochromatic pattern observed on the 747 landing-gear model made of ModelTech material.

Vibration of Jet Compressor Blades. Rupture of blades in turbines or compressors usually causes severe damage. Static stress analysis provides limited information related to the blade geometry only, not necessarily to the total behavior of the blade in its dynamic operating environment. Three-dimensional photoelasticity has contributed to blade design in terms of understanding static stress concentrations, while photoelastic coatings have been used directly on blades to measure stresses under dynamic laboratory conditions and in actual operation.

Blade testing on a shaker. Analysis of a blade mounted on an electrodynamic shaker and subjected to different vibratory conditions provides significant data concerning the dynamic stress concentrations at the blade-root junction and the effectiveness of the mount-

Fig. 6.3—Isochromatic fringes observed during the B-58 landing-gear test.

ing. In general, the location of the maximum stresses and the magnitudes and directions of the principal stresses are functions of vibratory conditions. This situation makes stress analysis with strain gages very difficult since the location of the maximum stress changes with frequency and amplitude. Also, strain-gage rosettes are required for this type of application; but because strain gradients are high, integration over the gage grid area introduces a significant error (on

Fig. 6.4—Model of the 747 main landing gear subjected to static test.

the nonconservative side). Because of these considerations strain gages are less than ideal for the task.

Vibration problems of this type can be analyzed quite simply with photoelastic coatings. The pattern is observed with a reflection polariscope equipped with a stroboscopic light that is synchronized with the blade motion by using either a capacitive or an inductive

Fig. 6.5—Compressor-blade analysis, showing residual strains produced by plastic
   deformations.

pickup. Slow-motion observation of the changing isochromatic patterns is obtained by stroboscopic phase shift. Using the plane-
reflection polariscope, the change in the position of isoclinic fringes
with vibration amplitude can be observed. The rotation of the direction of the stresses changes in the ratio of twisting to bending load
on the blade with increasing amplitude.

Blade testing in a compressor. The first stages of jet-engine compressors operate at relatively low temperatures (close to room temperature). Photoelastic coatings can therefore be used in these first
stages to determine the engine operating conditions under which
plastic deformations develop in the blades. Observation of the

Fig. 6.6—Static test on a missile structure.

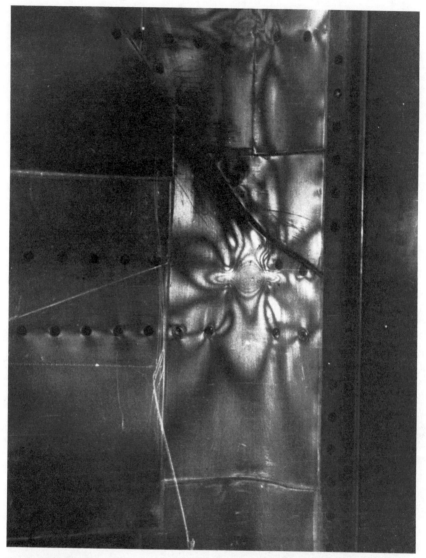

Fig. 6.7—Photoelastic coating reveals stress concentration at the wing-attachment stiffener.

residual-fringe patterns in the coatings is made after the engine stops. A typical residual-fringe pattern caused by plastic deformation is shown in Fig. 6.5. The location of the maximum-fringe order corresponds directly to the area where failures originate in this type of blade.

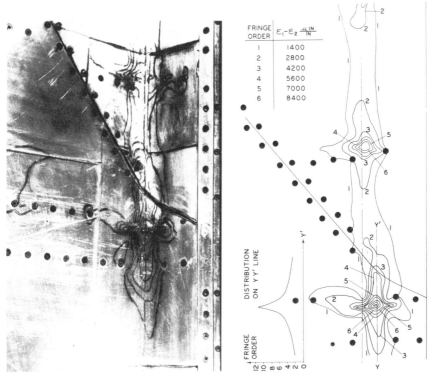

FRINGE ORDER	$\varepsilon_1 - \varepsilon_2 \frac{\mu\, IN}{IN}$
1	1400
2	2800
3	4200
4	5600
5	7000
6	8400

Fig. 6.8—Isochromatics traced directly on the coating and transposed into a graph of results.

Space Structures

Space structures are often similar to aircraft structures since both are designed for minimum weight. However, the space structures are often subjected to much larger loads because of the relatively short service-life requirements. Figure 6.6 is an overall view of a French missile subjected to static loads simulating certain flight conditions.[7] The isochromatic pattern in the vicinity of the wing attachment to the fuselage is shown in Fig. 6.7. Figure 6.8 is a plot of the fractional- and full-order fringes as traced with a grease pencil directly on the surface of the coating. When the fringe orders are low (e.g., one or two fringes) and a map of the isochromatics is required, the following procedure can be used:

1. Full fringe orders, as seen through a circular reflection polariscope, are traced on the coating with a grease pencil.

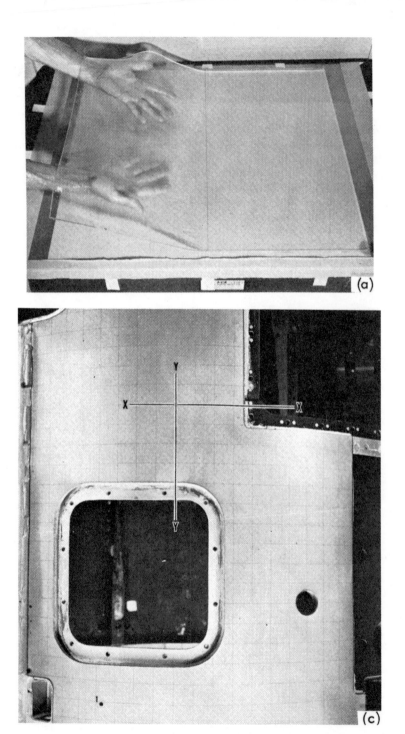

Fig. 6.9—Preparation of large coating: (a) lifting the contoured sheet from the mold, (b) placing the contoured sheet onto the bonding adhesive, and (c) and (d) location of measured points on the missile-guidance section.

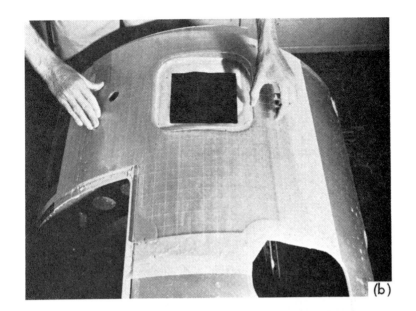

(b)

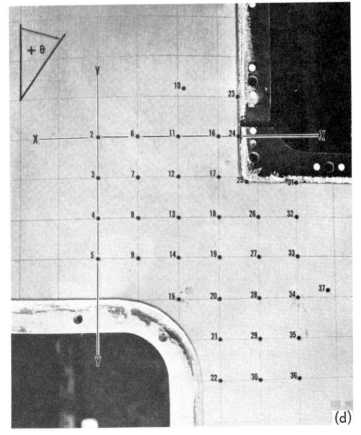

(d)

123

2. Fractional fringe orders (of any fraction) are also traced directly as seen in the circular polariscope, which is set to the fraction desired by rotation of the analyzer (the Tardy method). In areas where the isochromatic fringes fade (i.e., areas where the directions of the principal stresses are different from the direction of the polarizer axis), the polarizer, analyzer, and quarter-wave plates are rotated together until fringes reappear in the area of interest. This operation can be performed rapidly, and the results can be photographed to obtain permanent records.

While, usually, sheets of photoelastic coating 10×10 in. (250×250 mm) thick are applied to structures, in the case of large surfaces, larger sheets can be applied (see Fig. 6.9 and Plate C.6). The missile in these photographs was manufactured by the Douglas Company.[21] For this test series, a computer program was prepared to analyze the photoelastic data. Normal- and oblique-incidence readings of isochromatic and isoclinic parameters were fed into the computer to obtain the magnitudes, signs, and directions of the principal stresses.

6.2 PRESSURE VESSELS

Pressure vessels that are in regions of discontinuity around branch connections and the like are difficult to analyze. Three-dimensional photoelastic model analysis has been used extensively in such applications; however, photoelastic coatings have also been very useful, especially in areas where three-dimensional model analysis does not apply. These include problems due to yielding, cracks, welds, residual stresses, manufacturing tolerance errors, proof testing, and multiple-load analysis.

Figure 6.10 is a general view of a high-strength-steel rocket case subjected to a combination of axial load and internal pressure. A reinforcing band is welded around the central portion of the vessel. Plate C.7 shows a high fringe order that appeared at a very low load condition. The existence of a crack was suspected because the appearance of any such fringe pattern in areas having no geometrical discontinuity can indicate either a crack or local yielding. Fringe patterns due to cracks will disappear, or almost so, upon unloading the structure, while isochromatics due to yielding will remain essentially unaltered after unloading. Only the elastic contribution to the fringe pattern, usually one or two fringes, will be removed by unloading; the remainder of the fringe pattern will be permanent.

Photoelastic coatings were used to analyze both a scale model[46] and a full-sized rocket-case structure. A scale model was fabricated from the same material and by the same method as the actual full-

Fig. 6.10—Pressure vessel subjected to internal pressure, axial forces, and bending
    moment.

sized rocket case. Figure 6.11(a) is a representative fringe pattern
observed in tests of the scale-model rocket case. Figure 6.11(b)
shows a pattern observed on a fragment recovered after the comple-
tion of the burst test.

Quite often the maximum stresses in pressure vessels occur on interior surfaces and are inaccessible to the observer during the test.
However, a motion-picture or television camera can be placed inside
the vessel for observing and recording the fringe patterns. This technique was used, for example, on the Atlas missile. In a similar arrangement, cameras have been used to record the fringe patterns on
the exterior surface of a submerged submarine. A different technique can be employed when yielding is to be detected inside a pressure vessel. The procedure is as follows:

Fig. 6.11(a)—Photoelastic pattern observed on a model of a rocket case.

Fig. 6.11(b)—Coating exhibits yielding on a fragment.

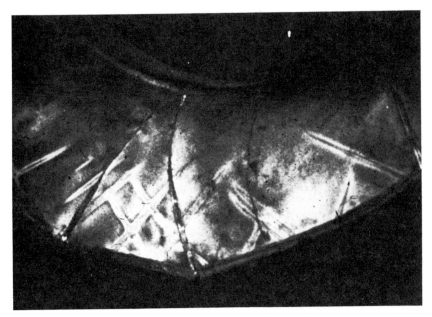

Fig. 6.12—Yielding detected on the interior surface of a pressure vessel; the photoelastic pattern was observed after the pressure was removed.

1. Coat the inside of the vessel in areas of interest.
2. Pressurize the vessel to a predetermined level; after emptying it, examine the coating for any permanent fringe patterns.
3. If no pattern is observed in step (b), reclose the vessel, pressurize it to a higher level, and reexamine the interior. Repeat this procedure until permanent patterns are observed or the maximum allowable pressure is reached.

Figure 6.12 shows such a permanent fringe pattern in a steel pressure vessel in the vicinity of a manhole opening. Figure 6.13 shows a pattern of thermal stresses developed in the solid-propellant grain that was cemented to the steel casing when it was cooled from casting temperature to room temperature. In the same test, stresses due to internal pressure were measured by using a heavy Plexiglas enclosure to enable observation.

6.3 AUTOMOTIVE

The automotive industry may become the largest user of photoelastic coatings. Design changes are customary, and the coating information can be used successfully for design improvements during a

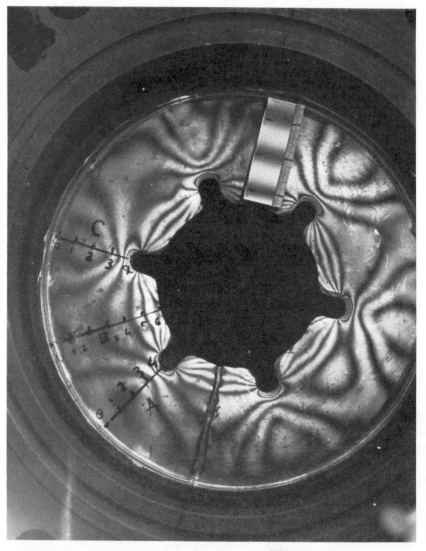

Fig. 6.13—Pattern of thermal stresses developed in the solid propellant grain.

design change.[47] The technique has been successfully applied to analyze stresses on engines, frames and suspension components, power train, etc.

Automobile Engine

Although automobile engines must be designed for "infinite" life (in terms of fatigue), weight savings are always important. These considerations make experimental stress analysis imperative. Problems caused by vibration, assembly stresses, thermal stresses, and fatigue are characteristic to automobile engines; and such problems cannot generally be solved with two- or three-dimensional photoelasticity because of difficulties in modeling.

The die-cast aluminum block of a 1961 model Rambler engine was analyzed by the photoelastic-coating technique.[47] Stresses were investigated for the following load conditions: (1) assembly stresses produced when the cast-iron cylinder head is bolted to the aluminum block, (2) vibratory stresses caused by engine operation, and (3) thermal stresses due to start-ups. During the analysis of assembly stresses it was found that many of the ribs in the engine block did not carry any load, while the threaded bosses holding the cylinder-head bolts were overstressed. Vibratory stresses were recorded with a high-speed camera during engine operation, but the observed stresses were very low in magnitude and could not account for service failures. It was noticed, however, that relatively high thermal stresses occurred during engine start-up.

The ribs that did not carry any load were cut away from the block with a milling machine, while the photoelastic coating was observed for evidence of stress redistribution. The net result of this effort was to reduce the engine weight by four pounds. The bosses that received the cylinder-head bolts were reshaped and slightly reinforced, producing a local stress reduction of 60 percent and eliminating engine-block failures. It was concluded that failure of the engine blocks had been caused by a combination of prestress, produced in assembling the cylinder head to the block, and high-stress, low-cycle fatigue damage due to thermal stresses during engine start-up. Figure 6.14 shows the aluminum engine block of the 1961 Rambler with the cast-iron cylinder head bolted in place. The fringe pattern caused by the assembly stresses in the original design is shown in Plate C.8, and that for the redesigned block, in Fig. 6.15. Note that the maximum fringe order in Fig. 6.15 is 60 percent lower than in Plate C.8. The bosses containing the assembly bolts are shown before and after redesign in Fig. 6.16.

Aluminum Steering-gear Housing

The Ross Gear aluminum steering-gear housing is used in different types of automotive vehicles and is subjected to correspondingly different modes of loading, depending on the particular mounting arrangement used. A simple method of studying the fatigue behavior of such housings was adopted by Ross Gear. The method, although empirical, is rapid and yields good results. The procedure consists of

Fig. 6.14—Rambler die-cast aluminum engine coated for analysis.

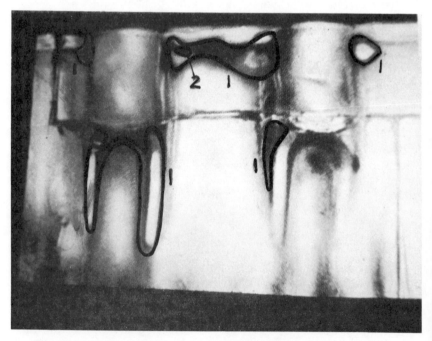

Fig. 6.15—Fringe pattern in the area shown on Plate C.8 after redesign.

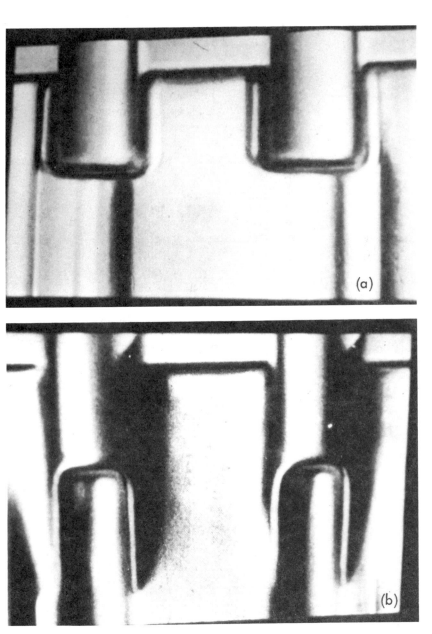

Fig. 6.16—(a) Boss area before, and (b) after modification.

applying a photoelastic coating to the steering-gear housing, fastening the housing to the vehicle or test fixture, and recording static assembly stresses after applying a few load cycles to align the assembly.

Fatigue cycling is then performed, and the dynamic-fringe pattern is recorded. The loads are scaled up to produce high-stress, low-cycle fatigue failures. The dynamic-strain amplitude is taken as the maximum strain in the cycle, and the design is rejected if the dynamic-fringe pattern indicates an amplitude greater than a specified value of strain. If the design is rejected, the housing is modified and tested against the same load level.

In the procedure followed by Ross Gear, the maximum dynamic-fringe order allowable corresponded to $\epsilon_1 - \epsilon_2 = P\,\mu\epsilon$. The value specified for P is obtained from the S/N curve for the material used in the housing. The assumption made here is that $(\epsilon_1 - \epsilon_2)_{max}$ is responsible for high-stress, low-cycle failures, which is valid if $\sigma_2 < 0$.

The preceding approach can be rather generally applied because wherever there is a dangerous situation in terms of $\epsilon_1 - \epsilon_2$, it is quite probable that a dangerous situation exists near this area in terms of ϵ_1. Consequently, when a modification is made to improve the distribution of $\epsilon_1 - \epsilon_2$ through redesign of the part, the distribution is then also improved for ϵ_1 in the redesigned area. Perhaps more important, $\epsilon_1 - \epsilon_2$, as inferred directly from the fringe pattern, is proportional to the maximum shear stress; and the maximum shear-stress criterion of failure is a close approximation to the Von Mises–Hencky criterion. Both the maximum shear stress and the Von Mises–Hencky criteria are good predictors of static yielding and fatigue failure of ductile materials under combined stresses. As a result, there are many cases in experimental stress analysis (often overlooked) for which the maximum shear stress is all that need be known; and separation of the principal stresses, by oblique incidence or otherwise, is unnecessary.

6.4 MINING AND CIVIL ENGINEERING

Mining (Rock Stresses)[48–50]

The photoelastic-coating method has been used in mines and found preferable to strain gages in some cases. Photoelastic coatings are particularly suitable where electrical instruments and wiring are not permissible in the mine shaft and where long-term measurements in very humid conditions are required.

In one application, internal (residual) stresses in rock strata were measured by bonding a photoelastic sheet to the rock and then making normal- and oblique-incidence photoelastic measurements after

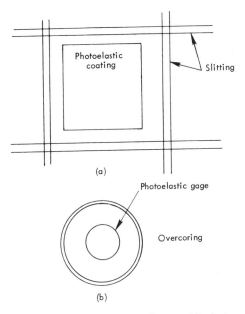

Fig. 6.17—Slitting of walls to relieve residual stresses.

cutting two parallel slots in the rock (above and below the photoelastic sheets). Following this, two additional slots were cut perpendicular to the first slots, and a second set of photoelastic measurements
was made [see Fig. 6.17(a)]. It is also possible to determine *in situ*
stresses in rock by using the "overcoring" procedure illustrated in
Fig. 6.17(b). In this application a small disk of the coating is bonded
to the rock, and a coring tool is used to cut a circular slot around
the photoelastic coating.[48]

Stresses in mines were also measured by inserting photoelastic
glass plugs in holes previously drilled in the rock to receive them.[49]
Any changes in the rock stresses (e.g., due to opening another shaft)
can thus be measured from the photoelastic pattern. Figure 6.18
shows the photoelastic pattern in such a plug during calibration tests
in a compression machine. As the ratio of the biaxial load changes,
the fringe pattern changes so that the relative magnitudes and directions of the principal stresses can be determined. Surface-bonded
photoelastic sheets can also be used for this type of application, but
the following precautions should be taken: (1) the rock surface
should be dried out and filled with epoxy, after which the epoxy
filler is cured with heat lamps prior to cementing the photoelastic
sheets in place; and (2) the coating should be covered with grease to
minimize bond weakening due to excessive moisture absorption.

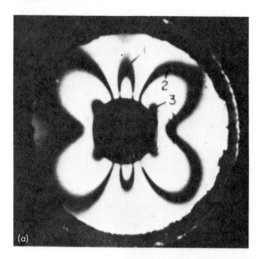

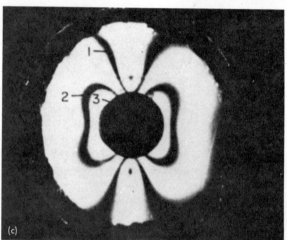
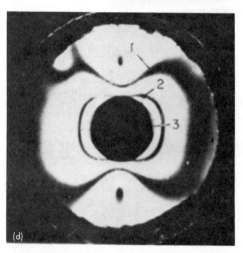
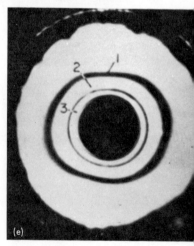

Fig. 6.18—Typical patterns observed on embedded photoelastic gages.

Bridges and Dam Locks[51-53]

Photoelastic coatings have been employed on bridges and dam locks to measure stresses in beams in the vicinity of weldments and rivets.   The test load was applied to a dam simply by filling it; a bridge was loaded by parking vehicles along its length (in one case, a bridge was loaded by pulling on a bridge tower with cables). Strain gages as well as photoelastic coatings could have been used in these studies; but, in several cases, the coatings revealed severe general yielding in weldment areas, which otherwise might have gone undetected.   For example, Fig. 6.19 shows the photoelastic pattern in a bridge assembly in the area where two I beams were cross-welded.

Silicone grease is applied to protect the coating from humidity, and when not in use the coating should be covered with a thin sheet of plywood to prevent mechanical damage. In the case of long-term exposure to the elements, moisture may channel along the metal-epoxy interface; here, large sheets of neoprene rubber can be used to cover the entire installation to protect the photoelastic sheet from becoming unbonded.

In situ measurements on bridges and similar large structures are often difficult and dangerous if the observer and the instruments must be near the observed area. In these cases a television camera can be used for remote observation while the photoelastic coating is

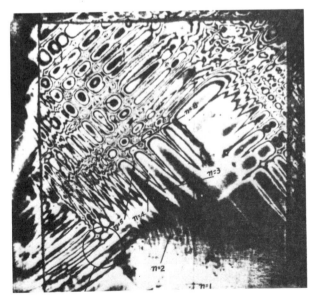

Fig. 6.19—Yielding observed in a welded region of a bridge structure.

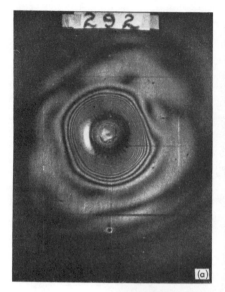

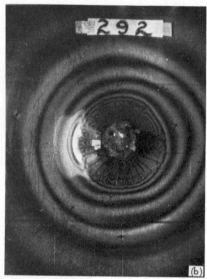

Fig. 6.20—Frames from a high-speed movie recording of a photoelastic pattern at 500,000 frames/s.

directly illuminated with a polarized-light source located at the coated area.

6.5 ORDNANCE

Armor Penetration by a Bullet[27]

Stresses due to impact, such as those caused by shocks, explosions, and bullets, have been investigated widely with the photoelastic-coating method. Figures 6.20(a), (b), and (c) show different isochromatic-fringe patterns in an armor plate being penetrated by a bullet. The transient patterns were recorded with a high-speed motion-picture camera (500,000 frames/s), and the polarized light source was an argon bomb. Electronic flash tubes of extremely high intensity and suitable for this purpose are now commercially available.[27]

Explosive Forming

Figure 6.21 is the photoelastic pattern observed on a coated aluminum plate after it was explosively formed. The fringe patterns revealed the flow of material during the very rapid forming process, and this study provided the insights necessary to improve the technique of explosive forming. In applications such as this one and armor penetration, the coating should be thin and capable of high elongations. The elongation capability of a polycarbonate coating is superior to epoxy coatings, and polyurethane-epoxy coatings are available for very large elongations.

Firearms

Several studies have been made to analyze the dynamic-stress histories of guns during firing. High-speed motion pictures (at 1000 to 20,000 frames/s) must be used for such studies. Because the strains are usually low, coatings of 0.1 in. (2.5 mm) thickness and the highest available K factor should be selected.

6.6 HOUSEHOLD APPLIANCES

Dishwasher Pump

Although photoelastic coatings have been used on many different types of domestic appliances, the following example was selected because the coating method provided a simple and inexpensive solution to an apparently difficult problem. A dishwasher pump of a particular design was failing (cracking) in normal kitchen service. Two alter-

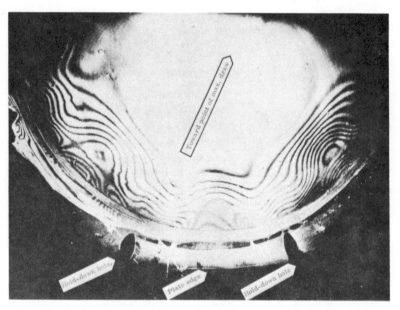

Fig. 6.21—Photoelastic pattern on an explosion-formed hemispherical shape, h_c = 0.048 in. (1.2 mm).

nate designs were developed in an attempt to strengthen the pump, but neither design eliminated the problem.

Elaborate studies of stresses due to cavitation, vibrations, and material analysis were unhelpful. Finally, photoelastic coating was applied to the two shells of the pump before assembly. The pump was assembled, and bolts were tightened with a torque wrench to the specified torque. This test, illustrated in Fig. 6.22(a), revealed very high stress concentrations around the first bolt as shown by the residual fringe patterns presented in Fig. 6.22(b). Starting up and pumping hot water introduced additional thermal stress, with the result that the pump failed in the same manner as in the field after several starts and stops. The cure was simple: alleviation of the static prestress around the first bolt by reducing the bolt torque to a tightness just sufficient to prevent leaks.

Washroom Sink

A new material was developed for the fabrication of a washroom sink by the hot-mold process. Several sinks were installed in a motel on a trial basis; but in a short time, cracks began to appear in the drain area. Photoelastic coatings were applied to the sink surfaces, and high stresses were observed due to cold-water/hot-water cycling. Plate C.9 shows a sharp isochromatic fringe coincident with the wa-

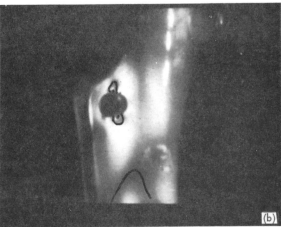

Fig. 6.22—(a) Setup for stress study, and (b) stress concentration due to assembly operation.

ter surface. Cuts were then made in the sink around the drain area to release any residual stresses that might be present. The sink was then reexamined and the results shown in Plate C.10 were obtained. Relief in the fringe pattern due to cutting indicated high residual stresses. As a result of these studies it was concluded that the field failures were caused by a combination of residual stresses and thermal fatigue.

Fan

A photoelastic coating was applied to the hub and blades of an axial-flow fan. The fan was dynamically balanced and then operated at normal rotational speed. Two tests were conducted to assess the merits of two different blade retainers. A reflection polariscope fitted with a stroboscopic light was used for the measurements, and synchronization of the stroboscope and fan was accomplished with a photoelectric cell. The discontinuous light intensity needed for the photocell signal was obtained from a piece of black tape on one area of the fan shaft. The following conditions were found:

1. Assembly stresses greatly exceeded dynamic stresses and produced plastic deformations in certain areas.
2. Centrifugal stresses were negligible.
3. Stress concentrations were almost totally absent in the blade fillet area, indicating excellent force transmission between the blade and hub during fan rotation.
4. One of the blade retainers created assembly stresses three times higher than the others.

Plate C.11 illustrates a large industrial blower with the coating applied. The isochromatic-fringe pattern due to assembly stresses near the fan shaft is shown in Plate C.12.

6.7 NONMETALLIC MATERIALS

Materials of Low Elastic Modulus

For stress measurements in materials such as rubber and cloth, coatings of low elastic modulus (below 20,000 psi or 140 MPa) and elongation capability of at least 20 percent should be used. Low-modulus coatings are required to minimize the reinforcing effect. These coatings can be applied in the form of either flat or contoured sheets. Although the coatings are low in strain-optical sensitivity (with K factors typically 10 to 20 percent of those for conventional coatings), they will usually provide high fringe orders when applied to soft materials because of their characteristically high elongations. The surface must be treated for adhesion before bonding coatings to some types of rubbers. In other aspects the technique is the same as for conventional coatings.

Automobile Tire. Tests were conducted to better understand the stress behavior of automobile tires. Tires were coated with photoelastic plastic on the outside rubber surface and, in some cases, directly on the cords after rubber removal. Because the strains were very high, thin coatings—usually 0.040 in. (1 mm) thick or less—were used for this application. Analysis was made for tires subjected to the following loading conditions: (1) internal pressure, (2) pressure plus an external vertical load, and (3) pressure plus an external load plus rotation of the tire. The coating is shown applied to the cord area in Plate C.13. Plate C.14 shows the tire subjected to a vertical load with the coating applied to the sidewall area.

Parachute Harness. A study was made of new garment types intended for use by pilots and designed in an effort to distribute the parachute-opening shock load over a larger portion of the body than is the case with presently used gear. Tests were conducted on two different types of flight suits to which photoelastic coatings had been applied: (1) "full-cover torso" and (2) "bandolier." One full-cover-torso suit and two bandolier suits were tested. Testing was accomplished by fitting the garments on dummies, then dropping them from a specially constructed drop tower. Straps of predetermined length were attached to the shoulders of the suits so that the dummies were permitted to free-fall approximately 25 ft (or 7.5 m) before the shoulder straps became taut and simulated the shock associated with parachute opening.

Test data were collected by use of a Fastax high-speed camera (Model WF-3) with a circular analyzer covering the lens. The camera was situated 13 ft (4 m) from the drop tower, and the field of view was trained on the area of initial impact with the taut shoulder straps. The test garment was illuminated by eight 750-W photoflood lamps installed behind a large sheet of polarizer/quarter-wave plate laminate. A fine wire was strung across the path of fall so that the dummy broke the wire in passing, triggering the camera operation.

Among other results of this test program, was the conclusion that most of the shock load was distributed in the upper portion of the full-cover-torso garment. With the bandolier garment, a greater portion of the load was transmitted to the legs (about two to three times as much as with the full-cover-torso garment).

Heterogeneous Materials

Materials in this category generally have elastic moduli somewhere between 5×10^5 and 2×10^6 psi (3500 and 14,000 MPa), depending upon the particular material and the direction of loading. Conventional photoelastic coatings are very satisfactory for use on any of these materials. When testing concrete, however, the surface should first be sealed with epoxy before applying the coating. When the surface preparation is complete, the reflective cement can be applied and the coating bonded in place in the usual manner.

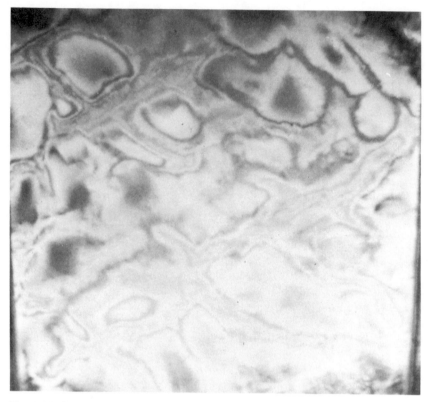

Fig. 6.23—Photoelastic pattern on a concrete specimen subjected to pure compression.

Concrete.[54] Several studies have been made to measure strain distributions in concrete. Very high, localized, strain gradients occur in
concrete under load because it is a heterogeneous material, composed
of mortar or cement and aggregate (stones). Since the cement has a
low elastic modulus, it deforms much more than the high-modulus
stones and induces severe strain gradients at the interfaces between
the cement and stones. Figure 6.23 shows the fringe pattern in a
coated concrete block loaded in compression in a testing machine.[54]
Figures 6.24(b) and 6.24(c) respectively give the strain distribution
$\epsilon_1 - \epsilon_2$ along line AB and the directions of the principal strains along
the same line. (See Fig. 6.24(a) for location of line AB.) Note that
the principal-strain directions are neither constant nor parallel to the
specimen axes, despite this being a prismatic specimen under a pure
compressive load. The deviations in principal-strain direction reflect
the local stress concentrations caused by differently shaped stones.

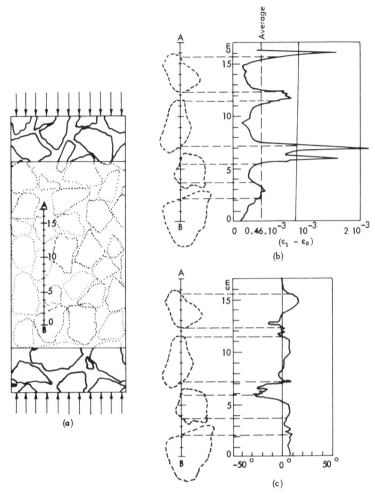

Fig. 6.24—(a) Location of the line AB across the outline of an aggregate, (b) strain distribution along AB, and (c) direction of maximum strain along AB.

Photoelastic coatings have also been used to detect and follow crack growth in concrete. A coated concrete model of a tunnel subjected to compressive loading is shown in Plate C.15. Nonuniform fringes (V-type) indicate the presence of a crack. Cracks can be detected by this method long before they can be observed with a microscope, even when it is focused continuously on the point where the crack will appear. The photoelastic coating method has also been used to measure creep in concrete as well as variations in stress distribution as a function of water content in a concrete specimen.

Fig. 6.25—Pattern observed on a wooden structure.

Epoxy–fiberglass and Wood.   Several tests have been performed using photoelastic coatings to assess the effect of the fiberglass reinforcement on the stress distribution in epoxy–fiberglass materials. Plate C.16 shows a coated epoxy–fiberglass plate containing a central hole and subjected to a tension load. The photoelastic pattern is generally similar to that observed in a coated plate of the same configuration made from a homogeneous material, but it can be seen that the reinforcement causes strong local departures from the normal fringe pattern.

Wood was also investigated with photoelastic coatings to determine the effects of knots, fiber direction, and moisture content on stress distribution. Figure 6.25 shows the photoelastic pattern on a coated wooden structure.

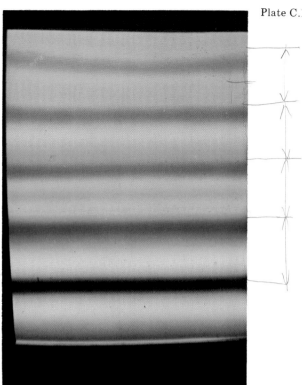

Plate C.1—Colored-fringe
sequence in a
linearly increas-
ing stress field.

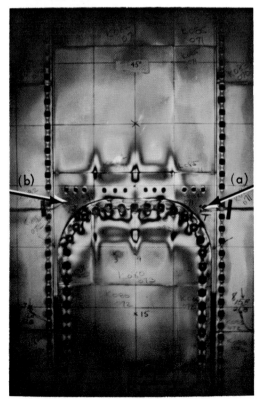

Plate C.2—Compression
load for C-141
wing-panel test.

Plate C.3—Residual strains in a buckled region on ribs.

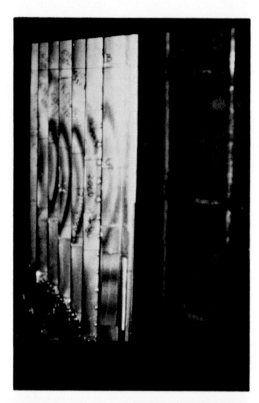

Plate C.4—Douglas DC-8 window-frame test.

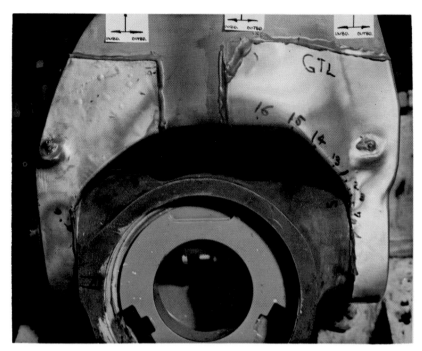

Plate C.5—Photoelastic pattern in a region of B-747 landing gear.

Plate C.6—Pattern on a large coating covering a spacer.

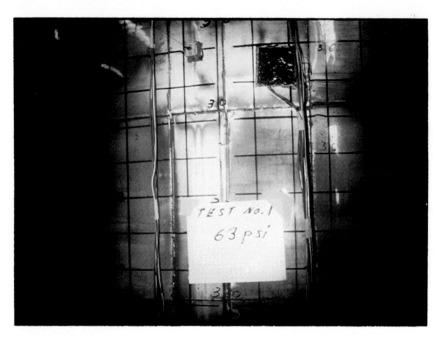

Plate C.7—Fringe pattern on a defect.

Plate C.8—Isochromatic pattern in highly stressed area of the engine shown in Fig. 6.14.

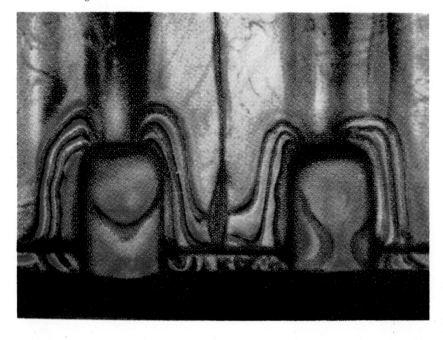

Plate C.9—Photoelastic pattern on the outside of the bowl shows the water line when the sink was filled with hot water.

Plate C.10—Residual stresses around the drain appear after relieving locked-in stress by cutting.

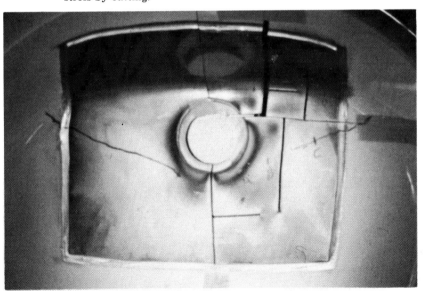

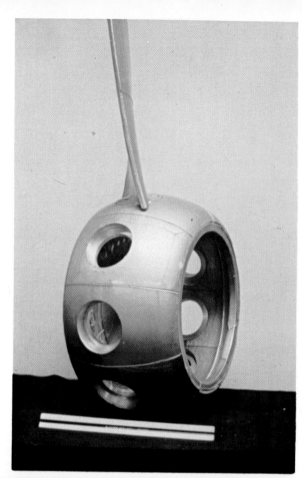

Plate C.11—Hub and blade with photo-elastic coating applied.

Plate C.12—Assembly strain pattern near the shaft.

Plate C.13—Photoelastic coating applied to an automotive tire.

Plate C.14—Photoelastic pattern observed on a tire subjected to vertical load and internal pressure.

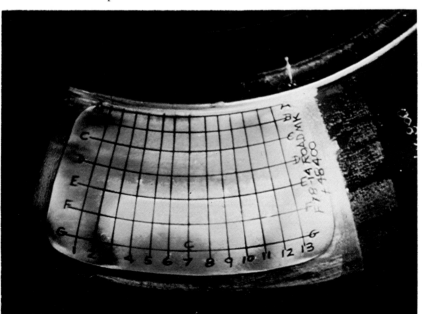

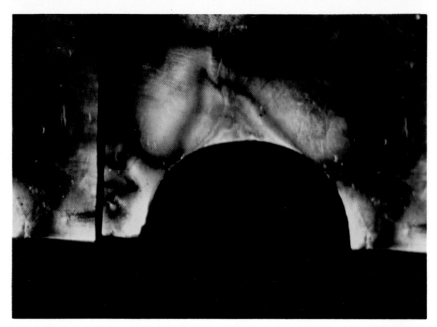

Plate C.15—Photoelastic coating applied to a concrete model reveals a crack.

Plate C.16—Photoelastic coating applied to a fiber-reinforced panel.

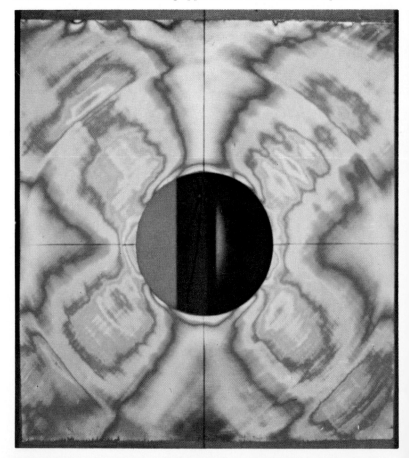

6.8 MEASUREMENT OF RESIDUAL STRESS

Photoelastic coatings can be used to measure residual stresses by any of the classical techniques in which the stress equilibrium is modified by removal of material or alteration of the geometry (e.g., changing a doubly connected geometry to a singly connected one). Although strain gages are extensively used for this purpose, a photo-elastic coating is occasionally preferable. This is particularly true when drastic full-field stress gradients may occur during material removal.

Composite-steel Cylinder[42]

Residual-stress measurements were made on a solid mild-steel cylinder 4 ft (1.2 m) in diameter and encased with a protective layer of stainless-steel weld metal approximately 1 in. (25 mm) thick over its entire surface. During welding, residual stresses developed in both the outer surface and the interior steel cylinder. Several sections were subsequently removed from the cylinder to measure the residual stresses. This description concerns two of those sections, one longitudinal and one transverse, as shown in Fig. 6.26. Photoelastic sheets, 0.120 in. (3 mm) thick, were bonded to specimens *A* and *B* (Fig. 6.26). The outer layer of weld metal was then machined off in shallow cuts to progressively relieve the stresses caused by welding. The strain distribution was measured after cutting each 0.040 in. (1 mm) of metal until the weld was completely removed.

Some of the results of these measurements are shown in Figs. 6.27–6.29. Figure 6.27(a) shows the variation of $\sigma_1 - \sigma_2$ along the boundary of the longitudinal section (specimen *B* in Fig. 6.26) and along the weld line. The amount of the weld removed at this stage was 1/4 in. (about 6 mm). The peak of the curve (at *a*) corresponds

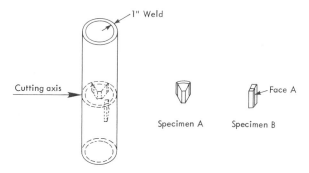

Fig. 6.26—Sections removed for residual-stress analysis.

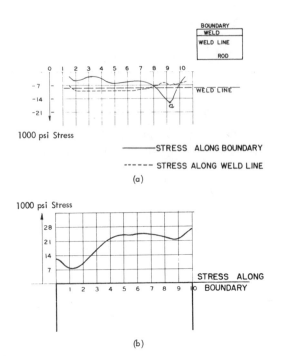

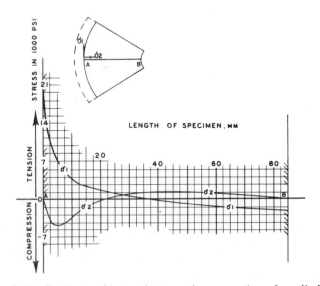

Fig. 6.27—(a) Stress distribution after partial removal of a weld, and (b) stress distribution after complete removal of a weld.

Fig. 6.28—Gradients of internal stresses in one section of a cylinder.

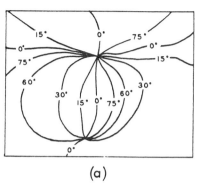 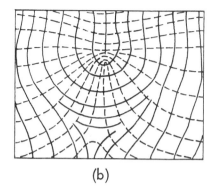

(a)   (b)

Fig. 6.29—(a) Isoclinic curves for the transverse section (specimen A) after complete removal of the weld, and (b) isostatics, or stress trajectories, derived from isoclinic curves.

exactly to the region where there was an interruption in the continuous weld.   Figure 6.27(b) represents the stress variation along the
boundary of the longitudinal section after the weld has been completely removed.   Figure 6.28 illustrates the stress distribution (σ_1
and σ_2) along line AB of the transverse section (specimen A) following complete removal of the weld.   The isostatic lines (stress trajectories) for the transverse section are shown in Fig. 6.29.

Another test was performed on a different transverse section cut
from the same cylinder.   In this case, residual stresses were determined by drilling holes in the weld and in the cylinder itself and then
measuring the stresses relaxed by the drilling operation.   The same
type of photoelastic plastic was used as on the longitudinal and transverse sections.   A typical variation of the difference of principal
strains along a line perpendicular to the hole axis is shown in Fig.
6.30.   The hole-drilling method is used typically as a point-per-point
technique, as opposed to the extensive sectioning used for full-field
analysis.

Aluminum Weldment

Two aluminum plates 0.25 in. (6 mm) thick were welded together at right angles.   Sheets of photoelastic plastic 0.076 in.
(1.9 mm) thick were bonded to the plates.   Figure 6.31(a) shows
three different cuts (aa', bb', and cc') made in the plates to relieve
the stresses introduced in their manufacture.   Then another cut dd'
was made to relieve the stresses caused by the weld.   All measurements for the principal directions and the difference in principal
stresses were made with a large-field polariscope.   An oblique-
incidence meter was used for determining the separate values of the

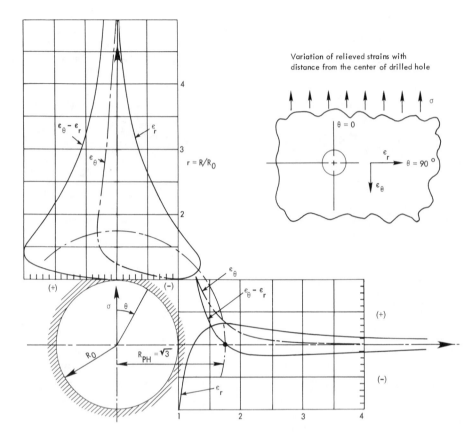

Fig. 6.30—Variation of relieved strains with distance from the center of a drilled hole.

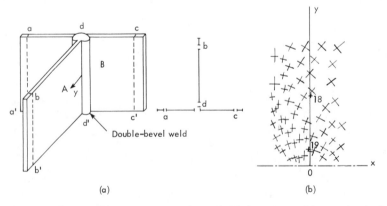

(a)   (b)

Fig. 6.31—(a) Plate-weldment test sample, and (b) directions of internal principal stresses in plate A.

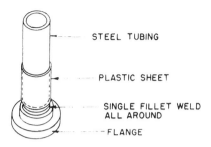

Fig. 6.32—Flange weldment.

principal stresses at specific points. Figure 6.31(b) shows the directions of the principal stresses at different points in the plate after cutting the plates along line dd'. Rotation of the principal-stress directions was caused by the residual stresses in the weld.

Tube Welded to a Flange

A steel tube 12 in. (300 mm) long of 2 in. ID and 3 in. OD (50 mm ID, 75 mm OD) was welded to a flange as shown in Fig. 6.32. A photoelastic coating was applied to the tube by the contoured-sheet technique. The weld-induced residual stresses were relaxed by cutting away the weld material to separate the flange from the tube. Figure 6.33 shows the variation in the directions and

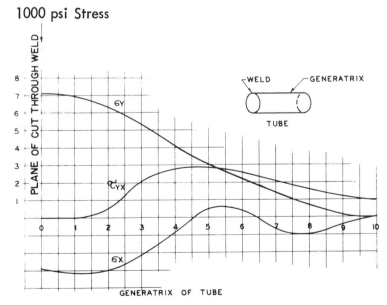

Fig. 6.33—Stress distribution along the generatrix of a tube.

magnitudes of the principal stresses along a longitudinal axis on the surface of the tube. These stresses are due to the residual reactions applied to the end of the tube by the weld material.

6.9 BIOMECHANICS

Photoelastic coatings have been used for a number of investigations in the field of biomechanics. For example, a human skull was coated and then subjected to shock loading to simulate blows with sharp and dull instruments. The coating and measuring techniques used for this study are identical to those for any other structure. The coated skull, ready for test, is shown in Fig. 6.34. In other examples, stress measurements were made on a coated femur subjected to compression in a testing machine, and on a coated human jaw compressed to simulate biting action.

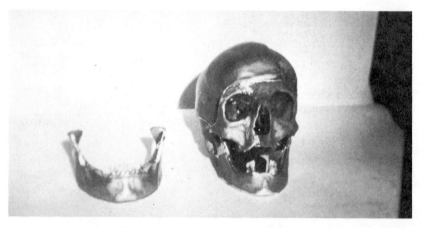

Fig. 6.34—Skull coated with photoelastic coating.

The photoelastic-coating technique has also found its way into the orthopedic branch of medicine, where different designs of leg and hip prostheses have been studied. In addition to these applications, forceps and other mechanical medical aids have been designed or improved through the use of photoelastic coatings.

6.10 NONDESTRUCTIVE TESTING[52,55,56]

In experimental stress analysis the emphasis is properly placed on the careful measurement of stresses for design improvement or for verification of the analytical assumptions used in the design. In nondestructive testing, on the other hand, there is more concern with reliability and quality assurance. Within the framework of the latter objective, the observation of surface strains with a photoelastic coating can be very useful.

Most of the currently used methods for nondestructive testing are directed toward the detection of defects in components while they are not subjected to service loads. These test methods unselectively indicate the presence of flaws, irrespective of whether such flaws are potentially damaging to the component performance. In contrast, surface strain observation (with the component loaded) will tend to detect only those defects or faults that cause stress problems and thus truly endanger the performance of the component. As a result of this characteristic, defects located in unstressed areas will not be detected, and unnecessary rejection of parts with such flaws will be avoided.

Assuming that the preceding generalizations are true, why is the photoelastic-coating technique for observing surface strains not more widely used in nondestructive testing? The answer to this question lies primarily with the conventional coating methods and the costs involved. However, practical techniques for applying the coating by spraying have recently been developed,[22] and this procedure offers an ideal application method where large areas are to be coated economically.

A surface defect such as a crack is very easily detected, even with an extremely thin photoelastic coating. If stresses are present in the region surrounding a crack, the plastic covering the crack is very highly strained, as indicated by considering the strain ϵ in the coating bridging the crack: $\epsilon = \Delta L/L_0$. In this equation, the initial width of the crack is L_0 (necessarily close to zero, or detection could have been made visually), and ΔL is the opened width of the crack as a result of applying loads to the part. From the equation it can be seen that the coating strain can be extremely high over the crack. An example of the fringe pattern due to a crack is shown in Fig. 6.35.

Subsurface defects, depending upon their nature, also influence the surface-strain distribution. The resulting surface-strain effects are demonstrated in Fig. 6.36 with a two-dimensional photoelastic representation of several common defects. This illustration shows the influence of: (1) a hole located immediately below the surface; (2) a hole at the center line of the plate; (3) internal cracks, one running parallel and one perpendicular to the surface of the plate and located in the middle plane; and (4) internal cracks parallel and perpendicular to the plate surface and located near the surface. For these demonstrations the plates are assumed to be subjected to uniaxial stresses in the direction of the axis of the photoelastic models.

The local increases in stress level on the surface, caused by vari-

Fig. 6.35—Pattern revealing a surface crack.

ous subsurface defects (expressed in percent of the stress increase over that in the area not affected by the local disturbance), are given in Table 6.1.[52] From Fig. 6.36 and Table 6.1 it can be seen that the local increase in surface stress is influenced by the shape of the defect and by the distance of the defect from the surface. Dangerous defects such as cracks perpendicular to the surface produce very sharp increases in stress which become readily apparent to the eye when the test object is photoelastically coated and loaded. On the other hand, defects that do not seriously threaten the integrity of the structure produce little or no effect.

Table 6.1—Influence of Subsurface Defects on Surface Stress

Position of Defect	Type of Defect	Local Increase in Surface Stress (in percent)
Near the surface	Crack parallel to the surface	7
	Crack perpendicular to the surface	50
	Circular cavity	33
Middle plane	Crack parallel to the surface	Negligible
	Crack perpendicular to the surface	25
	Circular cavity	5

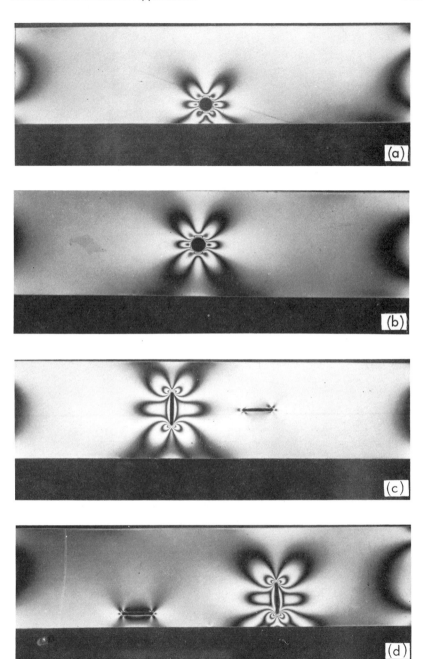

Fig. 6.36—Influence of internal defects on surface stresses: (a) circular cavity near the surface, (b) circular cavity in the middle plane, (c) internal cracks in the middle plane, and (d) internal cracks near the surface.

6.11 TRANSDUCERS

A photoelastic coating is a strain sensor and, as such, can be effectively employed in certain cases for transducer applications. In general, a photoelastic-coating transducer cannot compete with a strain-gage transducer for resolution and accuracy. The coated transducer may be advantageous, however, when any of the following attributes are demanded:

1. No contact between the transducer and the recorder.
2. No electrical instrumentation or circuitry permissible.
3. Extreme ruggedness necessary.
4. Low cost (expendable if appropriate).
5. Simple, "no-instrument" reading with no maintenance service.

In one example of a photoelastic-coating transducer, an automatic torquemeter was made by coating a power shaft. A polariscope fitted with a photocell was used for electronic recording of the birefringence. When the strain-induced birefringence is large, direct visual measurements of torque can be made from the rotating shaft. The

Fig. 6.37—Photoelastic transducer.

principal advantage of the photoelastic over the strain-gage torque-meter is elimination of the slip-ring requirement.

An inexpensive, low-accuracy photoelastic load cell is shown in Fig. 6.37. The transducer consists of a compression bar with a photoelastic insert (laminated between two polarizers). Loads are measured by noting the displacement of an isochromatic fringe along a calibrated scale. The accuracy of this transducer is about ±5 per-cent of full scale.

In addition, certain hand tools such as torque wrenches have been equipped with small reflective photoelastic strips having "fro-zen" fringe patterns. When the strip is covered with a circular polar-izer, the load or torque can be measured directly without the use of auxiliary instruments.

6.12 ALIGNMENT OF TESTING MACHINES[53]

The alignment of testing machines (and test specimens in the machines) represents a problem of critical importance in the study of material properties and structural behavior. It is remarkable that a small misalignment in load application can introduce a large change in the state of stress in the specimen. This characteristic is illustrated in Fig. 6.38, where the effects of small misalignments are demon-strated for a tension bar with a central hole. Any symmetrical body

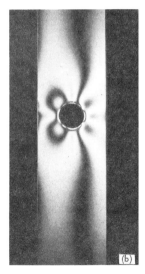

Fig. 6.38—Effects of load misalignment: (a) uniform distribution of stresses, (b) slightly eccentric loading, and (c) tension combined with twisting about the longitudinal axis.

that is loaded along an axis of symmetry should exhibit symmetrical stress and strain distributions.   This condition is shown in Fig. 6.38(a), where the photoelastic-coating pattern exhibits nearly perfect symmetry. As indicated near the top of the specimen, the tensile stresses are uniformly distributed across the bar.

In contrast, Fig. 6.38(b) shows the effect of slightly eccentric loading. Stresses along the right side of the specimen are 50 percent higher than those along the left side, and the photoelastic pattern is greatly distorted. In Fig. 6.38(c) torsion about the longitudinal axis is combined with the primary tensile loading. The presence of the torque significantly changes the principal-stress directions, as evidenced by the inclination of the isochromatic loops. Sensitive detection of extraneous torsional loading is also possible by observation of the isoclinic fringes because only a small torque is required to develop a substantial change in principal-stress directions.

Photoelastic coatings can be applied to the load-bearing members of testing machines as permanent installations.   Either continuous coatings or small photoelastic strain gages can be employed, depending upon the particular problem. Similarly, photoelastic coatings can be applied to representative test specimens to evaluate the quality of specimen alignment.

Photoelastic coatings can often provide a simple, direct method of calibrating testing machines. For the dynamic calibration of a fatigue-testing machine, a stroboscopic light is used to illuminate the vibrating coated part. When the flashing light is synchronous with the vibration, the calibration can be performed as for a static loading condition.

6.13 RESEARCH APPLICATIONS[54,55]

With the advent of practical photoelastic coatings, a new tool for many advanced research studies became available. Thus initial investigation proceeded on full-field plastic deformations, crack propagation, wave propagation, stress concentrations in the elastoplastic range, heterogeneous material behavior, and other phenomena previously difficult or impossible to analyze. Zero gage length, full-field capability, and the feasibility of using the method on any material made the coating technique ideally suited for this type of research. A few of many such basic studies are briefly mentioned here as examples.

Gerberich[56] has investigated the plastic-strain-energy density in cracked plates. Using photoelastic coatings, he observed that the degree of loading and strain hardening greatly affect the principal

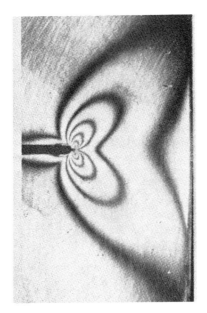

Fig. 6.39—Isochromatic of plastic deformations in 2024-0 aluminum near a
crack.

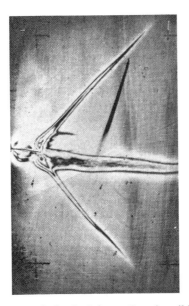

Fig. 6.40—Isochromatic of plastic deformations in mild steel near a crack.

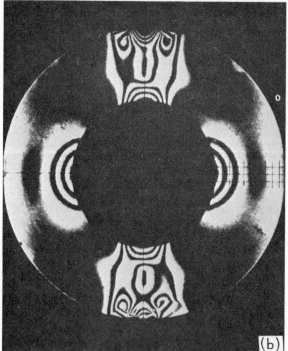

Fig. 6.41—Isochromatic patterns in analysis of elastoplastic behavior of a Dural ring: (a) loaded pattern, and (b) unloaded residual pattern.

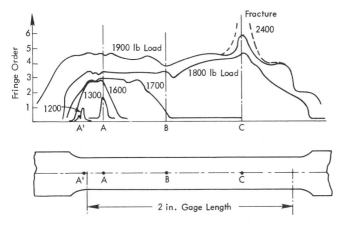

Fig. 6.42—Isochromatic-fringe order along the center line of a tension specimen as a function of load plastic deformations.

plastic-strain and energy-density distributions and that this effect is reflected in the amount of strain energy that can be plastically absorbed at a crack tip. Dixon and Visser[56] have also studied the propagation of cracks and Lueder's lines with photoelastic coatings. Figures 6.39 and 6.40 illustrate some of the results they obtained, showing the elastic and plastic stress fields in the local neighborhood of a crack tip.

Kawata,[6] Slot,[23] and others have studied stress concentrations in the elastoplastic ranges of deformation for different materials. Figure 6.41 shows the isochromatic-fringe patterns obtained in straining Dural specimens. The fringe pattern ,shown in Fig. 6.41(a) is for a load of 6000 lb on the ring, and the result in Fig. 6.41(b) is the residual pattern obtained after unloading the ring. This pattern represents the strain field due to plastic deformation of the ring. Several studies have also been performed on the plastic deformations of monocrystals; 99.9999 percent pure aluminum polycrystals were investigated under elastoplastic-deformation conditions. In addition, wave propagation in the elastic and elastoplastic ranges has been studied, as well as stress concentrations in anisotropic and heterogeneous materials.

The relative merits of the photoelastic coating and extensometer methods for determining the initial onset of local yielding have been investigated.[53]   Figure 6.42 shows the variation of isochromatic-fringe order along the center line of a mild-steel specimen subjected to several loads in the yield range. The first yielding observed photoelastically within the extensometer gage length occurred at a load of 1300 lb (5800 N), but the extensometer did not give its first indication of yielding until a load of 1650 lb (7300 N) was reached. This deviation of about 27 percent in determining the point of initial yield onset illustrates the error magnitudes that can be encountered in attempting to establish yield points with extensometers.

A conference held in Tallinn has indicated numerous applications of the photoelastic coating in testing and research in the Soviet Union.[58] In particular, Zacharov and Marov[42] have investigated the propagation of stress waves. Malishev and Zuckerman[44] have studied the effect of interaction of coating cement structure in the dynamic-wave-propagation condition.

Zacharov and Efremov[42] have reported applications of photo-elastic coatings to reinforced-concrete structures, with emphasis on crack detection and propagation. Also, Zacharov et al.[42] have reported research on use of interferometry on photoelastic coatings for the separation of principal stresses.

7
FUTURE DEVELOPMENTS

A number of current research and development projects that can be expected to improve and advance the photoelastic-coating technique are briefly reviewed in this chapter. Also noted here are some of the developments seriously needed to permit full utilization of the method's potential.

7.1 MATERIALS

Attempts are being made to improve the sensitivity (K factor) of photoelastic coatings, while reducing the modulus of elasticity. Commercially practicable developments in either or both of these directions would be very welcome, as one can judge from the content of the earlier chapters in this monograph. Materials are also under study for use as coatings at higher temperatures and for longer times than allowable with those that exist. Coatings currently in use deteriorate by carbonizing under severe temperature/time conditions. In addition, moisture absorption by photoelastic coatings is a problem that warrants attention. The absorbed moisture causes edge effects (parasitic birefringence) when viewed under normal incidence, and causes a uniform birefringence (with the appearance of biaxial strain) when observed under oblique incidence. To date, attempts to find materials that are insensitive to moisture have failed, and the common practice for minimizing this effect is to cover the coating with a protective compound such as silicone grease or RTV silicone rubber.

Several spray-on photoelastic-coating materials have been developed, but none is practicable for field use in quantitative stress analysis because of the problems in obtaining a uniform thickness of coating at least 0.040 in. (1 mm) thick. The coating can be built up by spraying successive layers, but each layer must be polymerized before the next is applied. This process normally takes several days and largely defeats the purpose of spraying. Moreover, a method for simple, easy measurement of the applied-coating thickness has not yet been found; such a method is particularly needed with sprayed coatings because the final thickness is both nonuniform and unknown. Notwithstanding these handicaps, the sprayed-coating method holds much promise for future development; it is presently suitable for crack detection in a part that can be loaded and for yield indication.

Because photoelastic coatings characteristically have very large thermal-expansion coefficients compared to metals, significant corrections in the coating indication may be necessary to eliminate the differential expansion effects when conducting thermal-stress analyses. Ideally, coating-expansion coefficients should match those of the structural metals, but the development of such coating materials is unlikely unless a form of glass can be made that is suitable for use as a coating.

The contoured-sheet method of photoelastic coating could benefit noticeably from the development of a material in which the polymerization could be arrested at the B stage and maintained there indefinitely until it was desirable that polymerization be completed. This feature would permit a laboratory to stock B staged sheets that were always ready for contouring. One approach to the problem (now under study) is to freeze the sheet at the appropriate point in the B stage and keep it frozen until needed. The sheet is then returned to room temperature before contouring, and subsequently applied in the normal fashion. Still another possibility is to use humidity as a curing agent before contouring.

Developmental effort is also needed in the direction of making it simpler and easier for the user to apply the coating. Progress has been made, for example, in providing premix kits (of both the adhesive and the sheet material) so that the user need not weigh or measure the resin and hardener, but merely mixes them. The development of a fast-curing cement such as Eastman 910 for photoelastic coatings would be very welcome. While a plasticized form of Eastman 910 cement is satisfactory for use on flat sheets of coating, it is very difficult to apply to contoured sheets.

7.2 INSTRUMENTS

As shown in Table 2.1, the maximum fringe order obtainable with photoelastic coatings when used for elastic stress analysis (the common case) is rather low. When working with very thin coatings or low strains, more sensitive instruments for improving the system resolution would be very desirable. There are two basic approaches to enhancing instrument sensitivity: increase the number of fringes optically by fringe multiplication,[59] or increase the resolution in measuring fractional fringe orders. Fringe multiplication up to 11X has been achieved by Day et al.,[60] but a practical industrial reflection polariscope incorporating this capability has yet to be developed. While Tardy and Babinet compensation methods for reading fractional fringe orders are in common use with reflection polariscopes, greater resolution would be advantageous when the maximum fringe order is less than 0.5, for example. New methods of compensation using a spinning-analyzer concept could be an improvement over currently used methods; and this type of instrumentation should be developed for strain measurements involving thin coatings and/or low strains. With use of this concept, sensitivity of 1/2000 fringe appears feasible.[61]

Although oblique-incidence systems are in general use to provide the additional measurement needed for separating the principal stresses, this method deserves refinement in several areas. For instance, the zero-order fringe is often difficult to recognize under oblique incidence because it is gray and lacking in sharpness. In addition, an oblique-incidence instrument is needed that is capable of fringe-order measurements in sharply concave surfaces. Data acquisition would be greatly simplified by a polariscope system capable of simultaneous oblique- and normal-incidence measurements, using two compensators, one for each direction of incidence.

Electronic systems have been used to improve and automate data acquisition from photoelastic coatings. Robert[62] has developed an instrument in which the analyzer is continually spun and the ellipticity of the emerging light measured. Direct photocell-readout methods are being investigated for obtaining accurate dynamic readings with this system. Computers are currently used for photoelastic-data reduction; there is no basic reason that continuous data from optical or electronic recording equipment cannot be fed directly to computers for data processing and retrieval in the same fashion as for multiple strain-gage installations.

Photographs of isochromatic-fringe patterns can be scanned and analyzed by a microdensitometer; an instrument of this type would contribute significantly to speeding and automating the data-reduction process. There is need also for an inexpensive camera system capable of simultaneously recording dynamic events at several locations in the manner of multichannel strain-gage systems.

7.3 METHODS

Normal-incidence isochromatic- and isoclinic-fringe patterns provide only two elements of information about the stress state in a coated object—the difference in principal strains (or stresses) and the directions of the principal axes. A third independent measurement is needed to determine the individual principal strains. Oblique-incidence fringe-order measurements are ordinarily used for this purpose, but the procedure is often inconvenient. Several other techniques have been studied, including the following:

1. Slitting: after slitting the coating along the observed principal-strain direction, the principal-strain magnitude can be measured directly under normal-incidence lighting because the resulting narrow strip is sensitive only to strain along its longitudinal axis.[16] If it is expected that the directions of the principal strains will vary, a hole drilled in the coating can provide both the magnitudes and directions of the principal strains. The hole method, however, requires calibration. Disadvantages of these methods include the fact that they are destructive and that initial birefringence may be introduced during machining, causing an error in the strain indication unless a no-load fringe-order measurement is made to correct the data.

2. Sum of the principal strains $\epsilon_1 + \epsilon_2$: if both the sum and the difference of the principal strains are known, the individual principal strains are directly calculable. The strain normal to the surface of the coating is proportional to the sum of the principal strains in the plane of the coating; as a result, $\epsilon_1 + \epsilon_2$ can be determined by measuring the coating thickness variation (isopachic pattern) produced by the stress state in the coating. Such measurements have been made by classical and holographic interferometry.

The separate principal strains have also been determined directly by measuring the absolute retardations.[63] All these methods merit further study as potentially practical solutions to a rather difficult problem.

Although the theory of thermal-stress analysis with photoelastic coatings has been firmly established by Zandman et al.[37] and field measurements have been successfully conducted, the method is both lengthy and difficult. The technique could benefit from the development of means for compensating for the initial birefringence caused by differential expansion between the structure and the coating (much as a dummy gage functions in a strain-gage circuit).

7.4 TRANSDUCERS

Many attempts have been made (with little commercial success) to employ photoelastic coatings as the sensing elements in transducers. Commonly, the transducer consists of a small strip of birefringent material, bonded to a rigid metal member and observed with a polariscope calibrated in the desired engineering units. Figure 6.37 shows an example of a transducer operating on this principle. In some instances, the measurement is made with a conventional reflection polariscope fitted with a compensator; in others, the compensator is attached directly to the birefringent strip and covered with a circular polarizer; in still others, the birefringent strip, which is covered with a circular polarizer, contains a frozen fringe pattern in such a manner that the isochromatics are shifted linearly as a function of the measured load or other variable. All these transducers that employ direct optical measurement are very low in sensitivity (perhaps 1 to 10 percent as sensitive as a strain-gage transducer). However, the relatively low cost of the transducers, their ease of reading, and the absence of electrical connections and electronic instrumentation are attractive features that make them particularly effective in mines and in civil engineering applications such as on bridges, dams, cables, etc. In the case of the torquemeter, the advantage arises from the elimination of slip-ring contacts.

The usual birefringent material employed in photoelastic transducers is epoxy. Nevertheless, glass offers certain distinct advantages for this purpose, particularly since (1) it can be stressed only in compression; (2) no machining is required; (3) bonding to the spring member of the transducer is unnecessary; and (4) its rigidity is beneficial. Within the framework of these constraints, glass offers the following advantages over plastics as the sensing material for transducers: higher strain-optical sensitivity, superior linearity and lower hysteresis, better stability, and insensitivity to humidity. In transducers employing glass as the sensing element, the glass is ordinarily inserted between two flats and prestressed in compression by the transducer structure, so that during operation the glass is always stressed in compression irrespective of sign reversal in the measured quantity.

Whenever the average strain (or stress) over a significant gage length is of interest, a photoelastic strain gage or stress gage can be used. Such a gage consists of a rectangle, square, or disk of photoelastic material that is bonded to the structure where the measurement is desired.

In conventional experimental stress analysis, the photoelastic stress or strain gage has only very limited application because the electrical strain gage, with a much smaller gage length, can accurately measure local strains and the gage does not significantly reinforce the test specimen in most cases. However, for certain types of transducer applications, the photoelastic gages may ultimately find a wide and effective usage. These applications would include inexpensive torque wrenches, cable transducers for prestressing concrete, overload indicators, yielding indicators, tension links, noncontacting torque-meters, fatigue and testing-machine alignment devices, and many others.

7.5 CONCLUSIONS

One of the major reasons why photoelasticity has never achieved the widespread usage of electrical strain gages is that the method, with its full-field ramifications, is difficult for many engineers and most technicians to understand. Because of the current industrial trend of increasingly relegating measurement tasks to technicians and away from graduate engineers, the technique of photoelasticity must be presented at the technician level. There is a growing need for new, simple devices to explain photoelasticity and to teach stress analysis in general at a more elementary level than in the past. The development of simpler, direct-reading instruments and simplified coating-application procedures can be expected to expand the use of the technique. The development of new coating materials, innovative instruments, electronic data processing, and improved methods will undoubtedly further this trend and will suggest many new applications. The photoelastic-coating method can be looked upon as a technique for experimental stress analysis which, although well established and basically mature, has by no means attained its full potential.

REFERENCES

1. Mesnager, M., "Sur la Determination Optique des Tensions Interieures dans les Solides a Trois Dimensions," *Comptes Rendus*, Paris, **190**, 1249 (1930).
2. Oppel, G., "Das polarisationsoptische Schichtverfaren zur Messung der Oberflachenspannungen am beanspruchten Bauteil ohne Model," *Zeitschrift des Vereines Deutscher Ingenieure*, **81**, 638 (1937).
3. Fleury, R., and Zandman, F., "Jauge d'Efforts Photoelastique," *Comptes Rendus*, Paris, **238**, 1559 (1954).
4. D'Agostino, J., Drucker, D. C., Liu, C. K., and Mylonas, C., "An Analysis of Plastic Behavior of Metals with Bonded Birefringent Plastics," *Proc. SESA*, XII (2), 115-122 (1955).
5. D'Agostino, J., Drucker, D. C., Liu, C. K., and Mylonas, C., "Epoxy Adhesives and Casting Resins as Photoelastic Plastics," *Proc. SESA*, XII (2), 123-128 (1955).
6. Kawata, K., "Analysis of Elastoplastic Behavior of Metals by Means of the Photoelastic Coating Method," *J. Sci. Res. Inst.*, Tokyo, **52**, 17-40 (1958).
7. Zandman, F., "Stress Analysis of a Guided Missile Tail Section with the Photoelastic Technique," *Proc. SESA*, **17** (2), 135-150 (1960).
8. _____. "Photoelastic Coating Test," *Nondestructive Testing Handbook*, ed. by R. McMaster, II, 53-1 to 53-39, Ronald Press, New York (1959).
9. Zandman, F., Redner, S., and Riegner, E. I., "Reinforcing Effect of Birefringent Coatings," *Exp. Mech.*, **2** (2), 55-64 (1962).
10. Dally, J. W., and Riley, W. F., *Experimental Stress Analysis*, McGraw-Hill, New York (1965).
11. Durelli, A. J., Phillips, E. A., and Tsao, C. H., *Introduction to the Theoretical and Experimental Analysis of Stress and Strain*, McGraw-Hill, New York (1958).
12. Holister, G. S., *Experimental Stress Analysis*, Cambridge University Press, Cambridge (1967).
13. Zandman, F., "Analyse des Contraintes par Vernis Photoelastiques," Groupement pour l'Avancement des Methodes d'Analyse des Contraintes, **2** (6), 1-12 (1955).
14. Dally, J. W., and Alfirevich, I., "Application of Birefringent Coatings to Glass-fiber-reinforced Plastics," *Exp. Mech.*, **9** (3), 97-102 (1969).

15. Redner, S., "New Oblique-incidence Method for Direct Photoelastic Measurement of Principal Strains," *Exp. Mech.*, 3 (3), 67-72 (1963).

16. O'Regan, R., "New Method for Determining Strain on the Surface of a Body with Photoelastic Coatings," *Exp. Mech.*, 5 (8), 241-246 (1965).

17. Perry, C. C., and Lissner, H. R., *The Strain Gage Primer*, 2nd ed., McGraw-Hill, New York (1962).

18. Leven, M. M., "Epoxy Resins for Photoelastic Use," *Proc. Int. Symp. Photoelasticity*, ed. by M. M. Frocht, Pergamon Press, New York, 145-168 (1963).

19. Ito, K., "New Model Materials for Photoelasticity and Photoplasticity," *Exp. Mech.*, 2 (12), 373-376 (1962).

20. Thomas, A. D., "Photoelastic Analysis and Model Fringe Value of Lexan Polycarbonate Resin," Technical Service Memorandum, General Electric Co. (Mar. 1962).

21. McIver, R. W., "Structural Test Applications Utilizing Large Continuous Photoelastic Coatings," *Exp. Mech.*, 5 (1), 19A-25A (2), 19A-26A (1965).

22. Instruction Manual for Bonding Flat and Contoured Photoelastic Sheets to Test Parts, Bull. IB-P-320, Photolastic, Inc., Malvern, Pa. (1962).

23. Slot, T., "Reflection Polariscope for Photography of Photoelastic Coatings," *Exp. Mech.*, 2 (2), 41-47 (Feb. 1962).

24. Frocht, M. M., *Photoelasticity*, Wiley, New York (1941).

25. Flynn, P. D., Feder, J. C., Gilbert, J. T., and Roll, A. A., "Some New Techniques for Dynamic Photoelasticity," *Proc. SESA*, 19 (1), 159-160 (1962).

26. Dally, J. W., Riley, W. F., and Durelli, A. J., "A Photoelastic Approach to Transient Stress Problems Employing Low Modulus Materials," *J. Appl. Mech.*, 81, 613-620 (1959).

27. Cole, A. A., Quinlan, J. F., and Zandman, F., "The Use of High Speed Photography and Photoelastic Coatings for the Determination of Dynamic Strains," *Proc. 5th Intl. Cong. High-Speed Photography*, 250-261 (1962).

28. Dally, J. W., and Riley, W. F., "Stress Wave Propagation in a Half Plane Due to a Transient Point Load," *Proc. 3rd Southeastern Conf. Theor. Appl. Mech.* (Mar. 1966).

29. Daniel, I. M., "Photoelastic Study of Crack Propagation in Composite Materials," *J Comp. Matl.*, 4 (2), 178-190 (1970).

30. Nickola, W. E., "Photoelastic Coatings on Flat Rotating Axisymmetrical Parts," *Exp. Mech.*, 4 (4), 99-109 (1964).

31. Dally, J. W., and Alfirevich, I., "Application of Birefringent Coatings to Glass-fiber-reinforced Plastics," *Exp. Mech.*, 9, (3), 97-102 (1969).

32. Post, D., and Zandman F., "Accuracy of Birefringent-coating Method for Coatings of Arbitrary Thickness," *Exp. Mech.*, 1 (1), 21-32 (Jan. 1961).

33. Duffy, J., "Effects of the Thickness of Birefringent Coatings," *Exp. Mech.*, 1 (3), 74-82 (1961).

34. Lee, T. C., Mylonas, C., and Duffy, J., "Thickness Effects in Birefringent Coatings with Radial Symmetry," *Exp. Mech.*, 1 (10), 134-142 (1961).

35. Duffy, J., and Mylonas, C., "An Experimental Study on the Thickness of Birefringent Coatings," *Proc. Int. Symp. Photoelasticity*, ed. by M. M. Frocht, Pergamon Press, New York, 27-42 (1963).

36. Theocaris, P. S., and Dafermos, K., "A Critical Review on the Thickness Effect of Birefringent Coatings," *Exp. Mech.*, 4 (9), 271–276 (1964).
37. Zandman, F., Redner, S., and Post, D., "Photoelastic Coating Analysis in Thermal Fields," *Exp. Mech.*, 3 (9), 215–221 (1963).
38. Alexandrov, A. J., and Achmetzanov, M. C., "Photoelastic Methods in Mechanics of Deforming Solids," Edit.: Nauka, Moscow, U.S.S.R. (1973).
39. Duffy, J., and Lee, T. C., "Measurement of Surface Strain by Means of Bonded Birefringent Strips," *Exp. Mech.*, 1 (9), 109–112 (1961).
40. Kolsky, H., *Stress Waves in Solids*, Clarendon Press, Oxford, 41–46 (1953).
41. Alexandrov, A. L., and Shandrov, L. G., "Wave Propagation in Photoelastic Coatings, Application to the Solution of Dynamic Problems," *Proc. 7th All-Union Conf. Photoelasticity*, Tallinn, U.S.S.R. 111–121 (1971).
42. Khesin, G. L., "Modeling of Dynamic, Thermal Stress and Static Problems Using the Photoelastic Method," Selected Papers, #73 Moskovskij Ordena Trudovogo Krasnogo Znameni Injenierno-Stroitelnyj Institut im. V.V. Kujbysheva, Moskva, USSR (1970).
43. Khesin, G. L., Filittov, I. G., Sacharov, V. N., and Marov, N. V.,"Some Problems in Application of Photoelastic Coatings in Dynamic Conditions," *Proc. 7th All-Union Conf. Photoelasticity*, Tallinn, U.S.S.R., 122–127 (1971).
44. Malishev, L. K., and Zuckerman, Y. N., "Application of Photoelastic Coatings to Dynamic Problems," *Proc. 7th All-Union Conf. Photoelasticity*, III, Tallinn, U.S.S.R., 128–135 (1971).
45. Riegner, E. I., and Scotese, A. E., "Use of Reinforced Epoxy Models to Design and Analyze Aircraft Structures," *J. Aircraft*, 8 (10), 813–817 (Oct. 1971).
46. Zandman, F., Watter, M., and Redner, S., "Stress Analysis of a Rocket Motor Case by the Birefringent-coating Method," *Exp. Mech.*, 2 (7), 215–221 (July 1962).
47. Zandman, F., and Maier, H. N., "New Techniques for Photoelastic Coatings," *Product Eng.* (July 1961).
48. Hast, N., *The Measurement of Rock Pressure in Mines*, Norstedt, Stockholm (1958).
49. Roberts, A., Hawkes, I., Williams, F. T., and Dhir, R. K., "A Laboratory Study of the Photoelastic Stress Meter," *Int. J. Rock Mech. Mining Sci.*, 1 (3), 441 (1964).
50. Hawkes, I., and Holister, G. S., "Photoelastic Techniques," in *Stress Analysis*, Ch. 12, ed. by O. C. Zienkiewicz and G. S. Holister, Wiley, New York (1965).
51. Zandman, F., "Photoelastic Coating Technique for Determining Stress Distribution in Welded Structures," *Weld. J. Res.* suppl. (May 1960).
52. Redner, S., "Nondestructive Testing Using the Photoelastic Coating Technique," *Materials Evaluation* (Nov. 1964).
53. Zandman, F., "Maximum Shear Stress Measurements and Determination of Initial Yielding by the Use of Photoelastic Coating Technique," *Symp. Shear and Torsion Testing*, ASTM STP 289 (1960).

54. Dantu, M. P., "Etude des Contraintes dans les Milieux Heterogenes, Application au Beton," *Annales de l'Institut Technique du Batiment et des Travaux Publique* (121), 55-77 (Jan. 1958).

55. Redner, S., "Prediction of Failures in Service by Photoelastic Method," *Advanced Testing Technique*, ASTM STP 476, 35-45 (1970).

56. Gerberich, W. W., "Plastic Strains and Energy Density in Cracked Plates," *Exp. Mech.*, 4 (11), 335-344 (Nov. 1964).

57. Dixon, J. R., and Visser, W., "An Investigation of the Elastic-Plastic Strain Distribution around Cracks in Various Sheet Materials," *Proc. Symp. Int. Photoelasticity*, ed. by M. M. Frocht, Pergamon Press, New York, 231-250 (1963).

58. "Photoelastic Coating Method," *Proc. 7th All-Union Conf. Photoelasticity*, Tallin, U.S.S.R., II, 123-196 (Nov. 1971).

59. Post, D., "Isochromatic Fringe Sharpening and Fringe Multiplication in Photoelasticity," *Proc. SESA*, 12 (2), 143-156 (1955).

60. Day, E. E., Kobayashi, A. S., and Larson, C. N., "Fringe Multiplication and Thickness Effects in Birefringent Coatings," *Proc. SESA*, 19 (1), 115 (1962).

61. Redner, S., "New Automated Polariscope System," *Exp. Mech.*, 14 (12), 486-491 (Dec. 1974).

62. Robert, A. J., "New Methods in Photoelasticity," *Exp. Mech.*, 7 (5), 224-232 (1967).

63. Favre, H., and Schuman, W., "A Photoelectric Interferometric Method to Determine Separately the Principal Stresses in Two-Dimensional States and Possible Application to Surface and Thermal Stresses," *Proc. Int. Symp. Photoelasticity*, ed. by M. M. Frocht, Pergamon Press, New York, 3-25 (1963).

INDEX